1 MONTH OF
FREE
READING

at

www.ForgottenBooks.com

By purchasing this book you are
eligible for one month membership to
ForgottenBooks.com, giving you
unlimited access to our entire
collection of over 1,000,000 titles via
our web site and mobile apps.

To claim your free month visit:

www.forgottenbooks.com/free976218

ISBN 978-0-260-84907-6
PIBN 10976218

FPL´68

Annual Report of Research at the Forest Products Laboratory

FOREWORD

Putting FPL Research to Work

There is a great gap in communications between the research laboratory and the production line, between an unknown something in a test tube and a new product in the hands of the consumer. It is generally recognized that the time-devouring, expensive task of developing new products, and the even more risky job of marketing them, rest with industry. Yet if progress is to be made, the research scientist must be concerned with surmounting the communications gap. This Annual Report is one means of surmounting that gap.

We in research, those in industry, and Government officials generally who influence FPL's destiny, are properly concerned with the role of a public laboratory in putting the results of its research to work. This is particularly true when one deals in such a broad area as forest products, which comprises 5 to 6 percent of our country's economy. Even more important is the current emphasis on improving the lot of our many citizens through research studies that include better and more economical housing and pollution abatement. As a public research laboratory, we are much aware of our role in influencing the long process from the scientist's bench to the salesman's shelf.

We have restudied our position with respect to all who use wood, the consumer, the wood products manufacturer, and the various intermediaries. For perspective we have started with a review of the Laboratory's history, beginning in 1910 and even before. We find that we have deep roots in the traditions of the Forest Service; our dedication is to "the greatest good for the greatest number" and "wise use is the best conservation." Research in the utilization of crops, of which timber is one, is widely pursued in the U.S. Department of Agriculture, of which we are a part. We find further a firm background of sound information and experience within the history of our own organization. We are led to marvel at the good judgment of the scientific pioneers of our Laboratory, who made great progress in the early years over a broad front of forest products utilization problems. For example, a John Newlin had a very practical approach to deriving working stresses for lumber. At the other extreme, a C. B. Norris developed highly refined mathematical analyses for the design of structural elements. Always we find emphasis on practical applications. A Harry Tiemann, who developed the basic concept of the fiber saturation point in drying, was also responsible for the first practical design of a dry kiln. And so we find, as we approach the 60th birthday of FPL, that we have some strong traditions to inspire and challenge us.

Today, if anything, our mission has been broadened. Briefly it is now defined as "research that leads to greater social and economic benefits through better utilization of timber resources." The barriers between research and application have been disappearing under the impact of a half dozen industry liaison committees, who meet with us once or twice a year; some 20 additional technical meetings at FPL every year; constructive consultation with over 3,000 technical visitors; some 90 cooperative studies with industry groups; and some 5,000 to 6,000 technical letters a year.

We are very conscious of the heavy responsibility that rests upon us. Only FPL, among the wood utilization research organizations of this country, can assume responsibility for supplying a steady flow of new basic information into the public pipeline. Only FPL has the responsibility and the capabilities for research programs on a broad national scale. Only FPL can plan and follow through on systematic, far-sighted, long-range programs of research, without interruptions due to changes of personnel and other disruptive factors. Only FPL has expertise in many different disciplines that have to do with wood use, and that can be drawn upon to solve the more complex interdisciplinary research problems that confront us today. Only at FPL is it practical to accumulate, maintain, and use for the attainment of national objectives so wide a variety of large, complex, expensive facilities essential to research, including an array of modern equipment and instrumentation, library facilities, and the like.

In spite of the growing complexity of our work, we hope to continue to provide leadership in forest products research around the world. We hope to be able to fulfill our mission in such areas as more efficient use of this Nation's declining timber supplies; more and better wood housing through more efficient and economical construction systems; ways of using forest and factory residues; greater product serviceability that will lead to better consumer satisfaction; and better and more efficient manufacturing processes that do not pollute the environment. This Annual Report presents some of the current research designed to reach our goals.

H O Fleischer

DIRECTOR

CONTENTS

FPL 1968:
Accent on Urgency

FPL's mission-oriented research assumed dramatic new dimensions in 1968 as its staff undertook tasks intimately associated with urgent national needs.

Foremost was a directive from the Secretary of Agriculture to help solve the critical housing problem in rural America. Out of this came plans and specifications for low-cost houses of wood construction, based upon FPL research and know-how, together with a manual on good construction practices and techniques for guidance of semiskilled rural workers. The houses were to be primarily for low-income rural residents. Coupled with this directive was a call to accelerate housing research.

Of comparable urgency was a second assignment: To help lessen the Nation's intensifying air and water pollution through research aimed at better utilization of wood harvesting and processing residues. This constituted recognition of the fact that disposal of woods and mill residues contributes materially to such pollution, and that FPL's research on ways to utilize residues needs to be stepped up and amplified to speed discovery of economic means of pollution abatement.

These assignments opened new opportunities to apply FPL's diversified skills and disciplines to many-faceted problems. The call for low-cost house plans, for example, involved not only engineering know-how but the pathologist's knowledge of decay hazards, the chemist's background in toxic preserva-

tives and treatments, and even architectural creativity in compact design. By year's end, three sets of plans and specifications had been drawn up for consideration by USDA agencies, notably the Farmers Home Administration and the Agricultural Research Service. And a contract was awarded by FHA in December, using one of FPL's plans, for the construction of the first house in Wisconsin—and one of the first in the Nation—to be financed for a low-income rural family under the Housing Act of 1968, enacted only a few months earlier by the Congress.

To put new emphasis on research related to pollution abatement, a research work unit on chemical engineering systems was established in the Division of Wood Fiber Products Research, and manpower recruited for it. The unit will do research on recovery of cellulose from pulp mill and paper mill effluents that contribute to pollution of streams. Work will also be done on the utilization of logging slash burned in the woods and sawdust and other refuse burned at mills—both responsible for air pollution in forested regions.

Reviews of background and current research status of both low-cost housing and pollution abatement are presented in separate sections of this report. This represents a departure from formats of past Annual Reports, reflecting the increasing mission orientation of research that cuts across traditional divisions of work and scientific disciplines.

Through snowy December, work was pushed on a new building for wood chemistry research. The structure will adjoin another new laboratory previously completed and in use.

A third separate section of this Annual Report deals with a similar broad subject, the evolution of strength evaluation and lumber grading from visual to mechanical techniques. It is a subject that has demanded much FPL time and attention for more than half a century.

International Activities

FPL's world-wide associations were reflected in activities with international organizations and foreign research institutions. Director H. O. Fleischer made two trips to Europe on research-related business. In July he met with top officials of the International Union of Forestry Research Organizations in Prague, Czechslovakia, as chairman of the Union's Section 41, on Forest Products. He also visited Rome to plan for a meeting of a wood-based panel products committee of the Food and Agriculture Organization of the United Nations, and Krefeld, Germany, to discuss with industry officials an FAO proposal for panel products pilot plant development in the tropics.

Dr. Fleischer returned to Rome in late October to preside as chairman over a meeting of the panel products committee. Object of the meeting was to compile and evaluate information designed to help developing Nations establish plywood, fiberboard, and particleboard industries. On that trip he also conferred with Yugoslavian officials on financing of forest products research at laboratories in Belgrade, Zagreb, and Ljubljana with socalled "counterpart" funds obtained from sale of surplus agricultural products.

Dr. John W. Rowe, FPL bark extractives authority, in January addressed an international products symposium at the University of the West Indies in Kingston, Jamaica. In July he spoke to an international symposium on the chemistry of natural products, sponsored by the International Union of Pure and Applied Chemistry, in London, England. Another FPL chemist, Dr. Roger M. Rowell, delivered a paper on wood carbohydrate chemistry at a British Chemical Society meeting in New Castle, England, while in that Nation for a year's study of carbohydrate degradation at the University of Birmingham.

Joseph A. Liska, chief of FPL's Division of Wood Engineering Research, participated in a meeting of the International Council for Building Research Studies and Documentation at Ottawa, Canada; he formerly chaired the organization's United States National Committee.

Dr. Margaret K. Seikel returned from a year's research on the molecular structure of hydrolyzable tannins at the Melbourne forest products laboratories of the Australian Commonwealth Scientific and Industrial Research Organization. Neil D. Nelson, FPL wood technologist, went to Australia's National

The colorful costume of his native Nigeria was worn by Oluwadare Awe, that Nation's Assistant Conservator of Forests, when he visited FPL June 12. Here he is shown with Director H. O. Fleischer inspecting oak wall paneling processed by a new FPL seasoning method, press drying. Mr. Awe had completed study for a bachelor's degree in forestry at Northern Arizona University, Flagstaff, assisted by the Agency for International Development.
M 134 890-5

University at Canberra for a year of study and research on a Fulbright scholarship.

National Affairs

It was another busy year for FPL participants in scientific gatherings throughout the country.

Proposals for revision of the American Lumber Standards for Softwood Lumber resulted in a number of meetings in which FPL specialists participated as advisers. Technical questions centered mainly on (1) size standards for green and dry lumber and (2) methods for assigning stresses and modulus of elasticity values to structural lumber. Historical background of this development is outlined elsewhere in this Annual Report.

FPL staff members participated in annual meetings of the Forest Products Research Society, the Technical Association of the Pulp and Paper Industry, the American Chemical Society, and numerous

Members of an industry committee representing the fiberboard, particleboard, and insulation board industries paused for this photograph during a meeting October 8 to 9 at FPL. Left to right, front row, are Charles Gray, Insulation Board Institute, Chicago; J. Mason Meyer, American Hardboard Association, Chicago; James Roberts, U.S. Gypsum Co., Des Plaines, Ill.; Dr. H. O. Fleischer, FPL Director; Robert LaCosse, Insulation Board Institute, Chicago; Dr. J. F. Saeman, PPL Associate Director; middle row, Dr. Robert A. Hann, FPL technologist; Tom Walton, Georgia Pacific Corp., Crossett, Ark.; Kenneth Peterson, American Hardboard Association, Chicago; back row, William Berg, Flintkote Co., New York; Harry M. Demaray, Timber Products Co., Medford, Oreg.; Malcolm Stehman, West Virginia Pulp and Paper Co., Tyrone, Pa.; George Matter, Weyerhaeuser Co., Longview, Wash.; Ralph L. Nelson, Conwed Corp., Cloquet, Minn.; Robert Dougherty, National Particleboard Association, Washington, D.C.; Bruce G. Heebink, Laboratory technologist; Wayne Lewis, Laboratory research engineer; and Charles Morschauser, National Particleboard Association, Washington, D.C.   M 135 488-8

other scientific organizations. Dr. Fleischer addressed the American Pulpwood Association's annual meeting at New York City in February on the potentials for whole-tree utilization, including branches, tops, stumps, and roots. Associate Director J. F. Saeman addressed the national meeting of the National Forest Products Association in Boca Raton, Fla.

Dr. Irving B. Sachs became president-elect of the Midwest Society of Electron Microscopists. Albert N. Foulger served as executive secretary of the Society of Wood Science and Technology. Other national meetings in which FPL staff members took part included those of the American Institute of Timber Construction, the American Wood-Preservers' Association, the National Woodwork Manufacturers Association, the American Society of Plant Physiologists, the Society of American Foresters, the Chemurgic Council, and the American Society of Agricultural Engineers.

Staff members also joined in a Gordon Research Conference on carbohydrate chemistry at Tilton, N.H.; several conferences and symposia sponsored by the Technical Association of the Pulp and Paper Industry; and a Packaging Management Conference of the Aerospace Industries Association.

Meetings at FPL

FPL was host to a number of organizations, both public and private, that convened here for conferences. Among these were the Technical Review Board, American Institute of Timber Construction; Subcommittee on Fiber Stresses, U.S.A. Standards Institute; Panel Products Industry Liaison Committee; Wax Subcommittee, American Petroleum Institute; Technical Committee, Southern Pine Association; Packaging Advisory Group, Department of Defense; Paint Task Force and Preservatives Subcommittee, National Woodwork Manufacturers Association; Committee T-4, Thermal Treatment of Poles, of American Wood-Preservers' Association; American Society for Testing and Materials; Committee on Technical Studies, National Forest Products Association; Joint Lumber-Paint Liaison Committee of National Forest Products Association and National Paint, Varnish, and Lacquer Association.

Liaison Committee for Fiber and Particle Panel Materials Producers; Research Liaison Committee, American Paper Institute—Technical Association of the Pulp and Paper Industry.

HOUSING
RESEARCH

In 1968, FPL research related to housing took on strong new emphasis as nationwide preparations by Government and industry to replace millions of substandard dwellings created demands for fresh concepts, techniques, and economies. The enormous requirements were dramatically brought into focus by Congress' enactment of the 1968 Housing and Urban Development Act, which defined the national goal as 26 million new housing units within 10 years.

Months before the legislation was enacted, however, FPL research, already responding to recognized housing needs, gained additional impetus from a directive of the Secretary of Agriculture to put Forest Service housing research to work on rural America's needs. The Secretary stressed the necessity of reversing the long-term population movement from country to city by providing better rural jobs, homes, and social conditions. This need was subsequently recognized by a clause of the Housing Act that specifically provided financial aid for low-income rural families. Before year's end, more than half a century of FPL structural research and know-how had been translated into plans and specifications for three low-cost housing designs for needy rural residents—plans that were promptly utilized by USDA's Farmers Home Administration to put the Housing Act to work for rural America.

And construction of at least one FPL design had begun—a two-bedroom house for a Winnebago Indian grandmother in central Wisconsin. The Farmers Home Administration financed the construction under provisions of the 1968 Housing Act for subsidized low-interest loans.

All three house plans specify similar construction, featuring the most effective use of conventional and locally available building materials. Two are one-story, ranch style; the third is an expandable modified Cape Cod, featuring dormitory sleeping accommodations upstairs for eight children. Two more plans are scheduled for release in 1969.

A major cost-reduction feature of the designs is the use of preservative-treated wood posts as a foundation. Plywood skirting can be used to enclose the crawl space. Floor, exterior walls, and ceiling are well insulated.

Among numerous other construction economies are a single-thickness plywood exterior wall covering instead of conventional sheathing and siding; a single-thickness plywood floor; and sparing use of

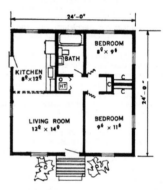

Floor plan and architect's rendering of FPL-designed house to be built for Mrs. Helen Cloud under terms of 1968 Housing Act for low-income rural families.

economy-type pine interior trim. Other economies are effected by careful planning of electrical, plumbing, and heating systems to hold wiring, fixtures, piping, and duct work to a minimum.

On the basis of mid-1968 prices, material and labor costs for the houses ranged from $4,000 to $7,000. Rising material prices later in the year forced these costs upward. Land, contractor's profit, and water and sewer installations were not included. Reasonableness of the estimates was confirmed near year's end by a construction bid of $7,022 for the two-bedroom house begun near Wisconsin Rapids, without carport but including a well and a septic-tank-dry-well installation and a concrete-block instead of a wood-post foundation.

As a further aid to builders of low-cost rural homes, a "Construction Manual, Low-Cost Homes of Wood for Rural America," was written. It is scheduled for publication in mid-1969 by the Superintendent of Documents, Washington, D.C. It is a well illustrated, step by step guide for building these

4

Twenty-one years after it was first erected at FPL to demonstrate a new concept called sandwich construction, this structure was dismantled for reassembly in the FPL Building Research Park. Floor, wall, and roof panels consist of paper-honeycomb cores and plywood, hardboard, or metal facings. Some panels were tested and found fully as strong as when the building was originally erected as a long-term exposure experiment.　M 135 468-5A

homes, starting with the staking out of the foundation and including the final finishing. It will be a particularly useful tool for those who wish to construct all or part of their own homes, for semi-skilled workers who may find useful employment in building these homes, and for the training of unskilled workers.

For many years USDA Agricuture Handbook 73,. "Wood-Frame House Construction," has been the authoritative source of information on all phases of wood house construction. Originally published in 1955, it has now been updated to include the most recent building materials and techniques. It is scheduled for publication in late 1969.

The new demands upon FPL personnel and facilities notwithstanding, research progress was made during 1968 in other areas directly affecting housing. A resume of these undertakings follows.

Post Foundations

The successful use of a wood foundation under an experimental FPL house for 29 years—at which time the house was moved to another site and the foundation dismantled for evaluation — demonstrated the practicality of this type of construction when fortified with preservatives to fend off decay and insect attack.

The foundation had by no means reached the end of its useful life, the evaluation disclosed. At the time it was dismantled, the wood foundation elements were giving excellent service and effectively meeting the requirements of use. The southern pine foundation planks that walled off the earth around the crawl space were in good condition. Above-ground portions of the Douglas-fir posts also remained sound; portions buried in soil averaged about 20 percent lower in compression strength than the tops, but part of this fall-off could be attributed to higher moisture content.

Experience with this foundation, as well as much broader research and experience with use of preservative-treated wood underground as fence posts, utility poles, and piling, was utilized to draw up recommended standards for low-cost pole and post house foundations for inclusion in Federal Specification TT-W-571 on wood preservation treating practices.

Thermal Insulation for Houses

Up-to-date information on the thermal insulating value of wood and wood-base materials used in housing was compiled from research and published in Research Paper FPL 86, "Thermal Insulation from Wood for Buildings: Effects of Moisture and Its Control." The report, updated from a 1948 FPL report, describes methods of estimating heat loss from buildings and compares fuel savings possible with various constructions.

Load-Sharing Concept Evaluated

The basic principles that govern socalled "load sharing" among parts of a multiple-member structural system—for example, floor joists connected with a subfloor—were established through analysis of the mechanics involved. Among fundamentals found to govern load sharing is the requirement that the stronger members must also be the stiffer, and the weaker ones the more flexible. True load sharing, it was concluded, can only exist where deflection of the structural system is possible. True load-sharing as presently conceived is therefore unlikely to exist in conventional joist-and-subfloor systems. Nevertheless, such systems no doubt are stronger than the weakest elements because of the strength contributed by the subfloor and the fastening system. The true gain in strength remains to be worked out experimentally for various construction systems, notably nailed and adhesive-bonded.

Low-Cost Window Unit

A simple window unit with a fixed glass section and a smaller screened opening for ventilation purposes was devised that is quickly installed in rough openings without headers. The unit consists of a simple lumber frame that serves as jambs, with the other edge routed to admit glazing. A wood exterior casing is screwed to the jamb and holds the glazing in place.

The assembled unit goes directly into the rough opening and is fastened with nails driven through the studs into the jambs. The exterior casing overlaps the house sheathing and siding previously installed flush with the rough opening.

Exposed faces of jambs and exterior casing are covered with special paper overlay, which hides knots and other imperfections of low-grade lumber and is an excellent paint base. Interior edges of jambs are flush with dry-wall or plaster and covered with simple molding.

Particleboard Weathering

The weathering characteristics of particleboard exposed outdoors at four sites in the United States were analyzed after 5 years. Boards bonded with urea resin were found to be flaking and failing. In most specimens there was a slight increase in thickness, but some had become a little thinner. Thickening was ascribed to a gradual release of springback with alternate wetting and drying; loss of thickness was believed due to weathering away of surface flakes. Some boards had been steam-stabilized at time of manufacture to prevent springback, and these were found to have swollen less, indicating that the steaming had effectively relieved stresses.

Insertion of the second unit in a mullion installation. No header or extra studs are required.   M 135 956

Fire-Retardant Particleboards

Several fire-retardant treatments contributed substantial resistance to flame spread on experimental particleboards used in an exploratory study of fireproofing methods. Satisfactory results were also obtained in terms of smoke reduction, strength, and dimensional stability of the boards. The experiments were confined to newly made boards. Among fire-retardant chemicals found effective were monoammonium phosphate and certain borates. Chemicals applied in solution were more effective than those applied dry.

Press-Dried Flooring

Experimental floorings of press-dried lumber were laid on the subfloor of an experimental house incorporating a number of innovative design features collectively called Nu-Frame. All were adhesive-bonded to a plywood underlayment. The house itself had been erected in the FPL Building Research Park a year earlier (see 1967 Annual Report).

Dimensional stability and wearing properties of the floorings will be observed over a number of years. Included are $\frac{3}{8}$-inch oak strips with and

6

Among various experimental floorings laid in a new experimental house was this press-dried oak veneer being put down by FPL Carpenter Clarence Rohovetz. The ⅛-inch strip flooring was bonded to a plywood underlayment with adhesive.　　M 134 806-7

without tongue and groove and ⅛-inch-thick red alder that had been pressed under relatively high pressure while being dried to improve its surface hardness. Press drying has been found to improve dimensional stability of wood under conditions that cause swelling and shrinking, so it is considered to be especially well suited for residential flooring.

Durable Natural Finishes

Low-cost chemical treatments that blunt the sun's wood-destroying ultraviolet rays show real promise of furnishing long-sought clear natural finishes that enhance and preserve the beauty of house siding, millwork, and other exterior parts of buildings.

The treatments contain inorganic salts, such as ammoniacal copper chromate, that absorb ultraviolet rays. Wood siding specimens treated with them are in good condition after 2 years on the Laboratory's test fences. Many clear film finishes break down within a year or two.

The treatments, applied as water solutions, penetrate the surface to form an insoluble inorganic precipitate which imparts photodegradation and fatigue resistance without obscuring the wood grain. The solutions cost about 50 cents a gallon and can be used alone or as a base for high-luster natural finishes, such as silicone resins that allow ultraviolet to pass through.

Research is continuing on these systems to extend service life and improve color. Cooperators are the National Forest Products Association and the National Paint, Varnish, and Lacquer Association.

STRENGTH EVALUATION OF STRUCTURAL LUMBER

The variability of wood's mechanical properties has created utilization problems that, despite many decades of sound research and numerous landmark accomplishments, still resist complete solution. Within recent years, the character and scope of the problems have been much more precisely defined and limited, and far more versatile tools for dealing with them are within reach. On the other hand, the gradual succession from old-growth to second-growth timber has accentuated the problems by introducing more variables even as increases in population and other factors have persuaded processors that timber of all species and qualities must be utilized much more prudently.

In general, the utilization problems center about one over-riding need: To make more efficient structural use of wood, especially in the form of lumber. Existing utilization methods, evolved out of much empirical research and experience, satisfy the need less with passing time. The methods, despite many refinements and improvements over the years, are still wasteful.

Visual Grading Requirements

These utilization methods are related to the system long used for grading lumber according to quality — which in this context means primarily strength. The grading system is based on visual inspection and general assumptions gained from research and experience regarding the effects on strength of visible characteristics, such as knots, pitch streaks, annual ring width, and the like. Object of the grading system is to segregate the better from the poorer material. This overcomes variability only in part; the penalty additionally imposed for safety reasons remains heavy. Engineers are furnished strength values for a given grade of a species that are purposely much lower than the average strength of that grade. This is necessary to all but eliminate the chance that exceptionally low-strength pieces will be used in ways that dangerously weaken the structure.

At its best, therefore, the visual grading system has shortcomings for segregating the strong from the weak. Human error can add to its shortcomings. Fortunately, means are becoming available to further refine it, supplement it, and eventually perhaps reduce it to an auxiliary status relatively free of human failings.

For all its shortcomings, however, visual grading remains indispensable over the near future. Only concentrated research effort can produce the information needed to evolve more precise methods of evaluating the strength of structural lumber. In fact, the new methods will undoubtedly build on the base of research that has brought visual grading to its present state of efficiency. A brief review of the evolution of visual grading and the technology behind it offers useful suggestions for future research and development.

Development of Size Standards

Until the middle of the 19th century, lumber needs were generally met from local timber sawed to builders' requirements. As forests receded from population centers and shipping distances lengthened, however, the need for standards arose. During the late 19th and early 20th century, plentiful old-growth timber yielded adequate supplies of high-quality lumber, hence the earliest standards were confined to definitions of size—width, thickness, and length.

Since size is directly related to strength, size standardization represented a first step toward a system of grading for uniformity. Even size standards, however, remained regional and conflicting until after World War I. In April 1919 an American Lumber Congress was convened to discuss the problem. The National Lumber Manufacturers Association was asked to investigate. Shortly thereafter NLMA sought the advice of FPL, instigating a broad study of lumber sizes then in use. In 1923, FPL published USDA Circular 296, "Standard Grading Specifications for Yard Lumber," and Circular 295, "Basic Grading Rules and Working Stresses for Structural Lumber."

Meanwhile, lumbermen asked the U.S. Department of Commerce to issue a national lumber size standard. After numerous meetings, the first softwood lumber size standard became effective July 1, 1924, as Simplified Practice Recommendation No. 16. A Central Committee on Lumber Standards, formed earlier by the industry, undertook its administration. FPL was often called in to advise. The standard was modified in 1929, 1939, and 1953. The 1929 modification put seasoning and moisture content effects on sizes under jurisdiction of regional associations issuing grading rules. The effect of this action was to be felt only after World War II demands for lumber brought much green lumber into use—and until agreement was reached in 1968 to write separate size requirements for green and dry lumber into the standards.

Regional Grading Rules

Grading rules as such have always been the province of regional lumber manufacturers' associations, and considerable variation results from species differences, competitive factors, and uses. Rules for structural timbers got a stronger technical foundation with the publication by FPL in 1934 of "Guide to the Grading of Structural Timbers and the Determination of Working Stresses," as USDA Miscellaneous Publication 185. Refinements eventually permitted assignment of bending and compressive stress ratings to certain grades of 2-inch and thicker dimension yard lumber commonly used for joists, rafters, studding, and other structural members of light-frame construction.

The basis for these grading rules is a massive collection of strength data produced during many years of research and analysis. The great number of commercial softwood species, the variability of wood both among and within species, and the lack of suitable mathematical and statistical tools greatly complicated the task of evaluating the strength of lumber. That a comprehensive publication on basic strength and engineering data on American commercial species could be produced as early as 1935 is a tribute to the pioneer FPL researchers whose ingenuity and brilliantly accurate intuitive judg-

ments resulted in USDA Technical Bulletin 479, "Strength and Related Properties of Woods Grown in the United States."

Evolution of Sampling Procedures

Typical of the problems confronting them was that of sampling a given species for test specimens. Statistical tools were virtually nonexistent and many forest stands were inaccessible. Instead, judgment of foresters and sawmillmen were relied on in selecting "typical" trees of a species for sampling and test. Despite financial and statistical limitations, the data developed proved reliable and served industry well.

Accumulated knowledge of interrelationships between strength and other physical properties of wood, notably specific gravity, led to improved sampling procedures for the latter property during the late 1950s. These resulted in the first comprehensive sampling of quality of standing timber on a regional basis, using specific gravity as the criterion. Beginning with the four major southern pines and continuing on commercial softwoods in 11 western States and Maine during the mid-sixties, sampling is now under way in Wisconsin on both softwoods and hardwoods (see section on Wood Quality Research). Many thousands of standing trees have been sampled for increment cores to get specific gravity values. The data have proved of enormous value to the lumber and plywood industry.

Meanwhile, advances in statistical sampling

9

Shown adjusting a deflection scale on an Engelmann spruce specimen to be tested in bending are engineer B. Alan Bendtsen and technician Bert Munthe.

theory have made it possible to get species strength data of known reliability that are based on a sampling technique encompassing the entire species growth range. The economy and value of the new procedure were demonstrated during the past two years with spruce pine and Engelmann spruce (see sections on Wood Engineering and Wood Quality Research).

The basic problem implicit in present visual grading procedures, however, remains: How can lumber be utilized more efficiently—that is, more closely in accord with the inherent capabilities of any given piece?

Mechanical Stress Rating

In recent years, so-called stress-rating machines that test each piece have been introduced by industry. At low loads that do not damage the lumber, deflection is measured and a modulus of elasticity value is obtained for each piece that can be correlated with the strength properties.

Principles of Nondestructive Testing

FPL research in this area of nondestructive testing continues to emphasize development and elucidation of principles involving the relationships between physical and mechanical properties. It is recognized that visual grading for strength evaluation is in its way a kind of nondestructive testing too. A recent analysis of bending test results given in seven independent studies recorded in the literature shows how several measurable variables compare as grading criteria, singly and in combination. All four variables used—modulus of elasticity measured over full span (MOE), visually estimated strength ratio (ESR), specific gravity (G), and bending strength of small, clear, straight-grained specimens (CMOR)—are familiar. There is little evidence, however, that they have ever been studied collectively.

Results of the collective study are given in table 1 in terms of the impact of the four variables, and various combinations of them, upon the coefficient of determination, which is an index of the amount of knowledge each variable gives about bending strength. Some strikingly consistent trends are apparent:

1. There is no clear preference between ESR and MOE measured over full span as an indicator of bending strength.

2. ESR plus MOE always is some 10 percent higher than either alone. (A very similar improvement affecting tensile strength has also been noted.)

3. G plus ESR gives a similar but smaller improvement in the coefficient of determination.

4. Inclusion of CMOR with the sum of ESR and MOE is not helpful in accounting for the impact of these variables on bending strength.

These findings point the way for further re-

Table 1.—Coefficients of determination for bending strength related to several variables

Study No.	Independent variables				
	MOE	ESR	ESR + MOE	ESR + G	CMOR + ESR + MOE
1	0.42	0.44	0.59	0.54	0.59
2	.52	.60	.75	.69	.75
3	.35	.20	.45	.26	.45
4	.27	.21	.37	.39	.37
5	.67	.67	.78		.79
6	.49	.60	.69		.70
7	.42	.29	.48		.56

search on nondestructive testing methods that employ such variables. The combination of estimated strength ratio—an application of visual grading principles—and modulus of elasticity clearly indicates, for example, better accountability than a nondestructive test based on MOE alone for determining stress ratings of beams, joists, rafters, and other bending members. This supplies some tangible evidence that methods involving MOE measurement and visual restrictions, sometimes currently used, can provide much more reliable stress ratings. It is toward this objective that FPL research on the strength evaluation of structural lumber is being directed.

Strength Predictors Needed

The research area is broad. There are numerous characteristics of wood that can be measured and that can serve as indicators of mechanical properties—without destroying the piece being tested. Examples are, besides the variables discussed above, weight, moisture content, sound conductivity, growth rate, percentage of latewood, hardness, and vibration. Means of studying wood strength properties include, besides conventional stressing machines, studies with strain-sensitive coatings, x-rays, gamma rays, and beta rays.

There remains much to be learned—essentially all of it relating to strength properties and to the way design can proceed where precisely graded lumber will be used. Modulus of elasticity is perhaps the best understood of the mechanical properties involved. Further research, therefore, should focus on predictors of the several strength properties. Such predictors need to be isolated, and methods developed for measuring them.

Work along these lines is under way, at FPL and elsewhere. Perhaps one of the most urgently needed developments is to intensify the research studies wherever they are being done, and to speed their results into use.

POLLUTION ABATEMENT THROUGH UTILIZATION ADVANCES

By definition, pollution makes the environment unclean and impure. Woodpulp mills deposit wood residues and byproducts of chemical pulping in streams or the atmosphere. In some forest regions air is polluted by the burning of logging slash and sawmill residues. Even the disposal of cast-off wood products by burning at municipal dumps creates pollution. And in the larger sense, forests themselves may be said to have been polluted by centuries of cutting out the best and leaving the rest—ultimately creating forest ghettos crowded with small, stunted, and cull trees of extremely limited use potential.

The common denominator in all these different kinds of pollution is incomplete utilization. A third of the fiber in every tree is left in the woods as stumps, large limbs, and tops, which are frequently burned for sanitation, to reduce fire hazard, or to improve appearance. Commercial chemical pulping processes convert to usable fiber about half the substance in each pulpwood bolt. Upwards of a quarter of each sawlog is discarded as sawdust, bark, slabs, and edgings at many mills, sometimes to feed smoking refuse burners. Forest slums, mainly hardwood, are mute testimonials to past ravishment and neglect; although hardwood growth exceeds drain by 75 million cords a year, much of it has little use at present.

A long-standing objective of forest products research has been to reduce waste. In recent years mounting public outcries against pollution have put new urgency into this research. Excellent progress has been made in some areas. Many sawmills now sell chipped slabs and edgings and sawdust to pulp mills. Once-discarded wood also goes into millions of square feet of fiberboards and particleboards. Species and logs of qualities formerly rejected are routinely sawn into lumber and plywood.

Much more progress appears imminent on the basis of recent research. Among wood processes that offer more complete fiber recovery—and correspondingly less polluting waste—are two new or improved FPL processes, polysulfide and magnesium-base bisulfite pulping. Chemical recovery systems engineered for both appear efficient. Commercial pilot-scale development is now considered feasible.

More efficient log breakdown systems are also clearly discernible from recent research. A new FPL design for a circular saw appears capable of saving $\frac{3}{32}$ or $\frac{1}{8}$ inch of kerf—more lumber, less sawdust. New experiments in wood slicing hold promise of cutting flitches up to an inch thick wholly without sawdust.

Among recent FPL research developments pointing the way to utilization of more low-quality timber is press drying. Not only does this process remove moisture from green wood far faster; it also holds the wood flat and contributes dimensional sta-

11

Residue utilization, 1952 and 1962, in the United States. Each unit is 1 million cubic feet.
M 128 838

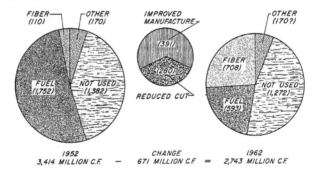

| 1952 | CHANGE | 1962 |
| 3,414 MILLION C.F. — | 671 MILLION C.F. = | 2,743 MILLION C.F. |

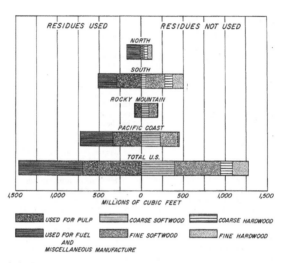

RESIDUES USED   RESIDUES NOT USED

NORTH

SOUTH

ROCKY MOUNTAIN

PACIFIC COAST

TOTAL U.S.

1,500   1,000   500   0   500   1,000   1,500
MILLIONS OF CUBIC FEET

USED FOR PULP   COARSE SOFTWOOD   COARSE HARDWOOD

USED FOR FUEL   FINE SOFTWOOD   FINE HARDWOOD
AND
MISCELLANEOUS MANUFACTURE

Distribution and utilization of residues by regions in the United States. M 128 837

12

bility to the dried products. FPL's componentized Nu-Frame house design calls for more low-grade lumber than conventional construction. Research on the mechanical properties of fiberboards and particleboards opens the way to structural load-bearing uses for these residue-constituted materials. Chemical research is facilitating the feeding of wood as roughage to farm livestock and may succeed in modifying wood enough at low cost to enhance its digestibility.

Over the longer term, research that is pointed toward more complete utilization of wood in all forms has even greater significance for pollution abatement. Work is under way on the characterization of the fibers in limbwood and the tops of trees now customarily burned as logging slash. Loblolly pine top-wood has been analyzed to get data on the fiber characteristics compared with those of juvenile wood laid down when the tree was young. Some 13.5 million fiber measurements on the wood of 17 trees were analyzed in a study of the effect of tree age on average cross-sectional dimensions of fibers.

Fiber characterization has implications other than for pulping. Studies have shown that the gelatinous fibers found in so-called tension wood of leaning hardwood trees may be more fully digested by ruminants than is normal wood. Tension wood is common in some cull timber.

Bark, on the other hand, besides containing machine-damaging dirt and grit, offers little promise of utility as a source of fiber. One practical disposal course therefore seems to be to develop methods of debarking logs in the woods and scattering the bark on the land for soil conditioning purposes.

There is firm evidence that residues of sawmills, planing mills, and veneer mills can win wider use if byproduct requirements are established. This can sometimes be done by modifying processing machinery to produce residues of size and form more suitable for particleboard to reduce resin binder requirements, and for pulping to avoid plugging of conveyors and digesters. With such objectives, sawtooth designs have been modified to produce coarser sawdust, and planers experimentally redesigned to produce more usable shavings (see section on Solid Wood Products Research).

Chip Deterioration in Storage

The problems of residue utilization do not end with conversion to usable form. Transport and storage requirements also pose complex technological and economic questions. Forest Service research is probing the possibilities of pipeline transport of chipped harvesting and thinning residues. An FPL task force is at work on deterioration of chips in storage piles that can cause losses of 5 to 10 percent a year and sometimes results in spontaneous combustion, with much greater losses.

Chemical and biological causes of chip deterioration are being investigated on a laboratory scale to

This insulated chamber holds 100 cubic feet of chips for study of deterioration under conditions simulating those in a commercial pile of wood chips.　　　M 135 627-12

Dr. Edward L. Springer operates tissue respirometer for studying rates of chemical and microbiological oxidation of gram-quantities of wood.　　　M 135 627-3

get needed basic information. Pathological studies of fresh Douglas-fir chips have uncovered a number of thermophyllic fungi and bacteria that are being examined for their destructive and heat-producing capacity. Means of rapidly evaluating fungicides

for use in spraying chips as they are piled are being sought. Chips are being stored under water and in nitrogen-carbon dioxide atmosphere to learn whether anaerobic micro-organisms can attack wood under these conditions. The effectiveness of thiamine depletion to prevent micro-organism attack is being evaluated.

Deterioration of chips is under study in insulated chambers that simulate pile conditions of oxidation and temperature rise. Oxygen consumption of wood particles is being measured with respirometers designed for use with living tissues, to learn about mechanisms of deterioration.

Five western pulp and paper mills are cooperating in this research.

Some Utilization Problems

More efficient utilization of harvesting and processing residues, with its companion benefits of pollution abatement, is largely dependent on the development of more effective technology through research. Among problems requiring further research are:

1. Economical techniques for handling small-size residues of variable form.

2. Knowledge of the range of quality of some residues, notably branch wood and other harvesting offal.

3. Effects of variable residue characteristics on product quality.

4. Drying technology for a variety of small wood forms.

5. Better log breakdown technology to reduce or eliminate saw kerf, get maximum yield from logs of all qualities through mill automation (see section on Wood Quality Research), and devise techniques to produce wood sizes and shapes more closely related to end-product requirements.

6. Effective control of biological and chemical causes of wood deterioration in storage, especially outdoor piles of chips.

7. More efficient chemical pulping processes to reduce loss of wood substance, especially the fibrous material.

8. Low-cost delignification treatments that can convert to digestible stockfeed some 15 million tons of coarse and fine wood residues that annually accumulate at processing mills.

9. Uses for residual solubilized carbohydrates at pulp mills.

10. New pulping procedures using such materials as oxygen to eliminate air pollutants like the sulfur compounds formed in kraft and sulfite pulping reactions.

11. More efficient reprocessing of fiber from used paper; some 30 million tons accumulate annually, only 10 million tons of which are now reprocessed.

13

SOLID WOOD PRODUCTS RESEARCH

The importance to consumers of products that are made of wood well seasoned and machined, that are fabricated with reliable adhesives, thoroughly protected from decay, fire, and other service hazards, and economically produced from wood of all qualities and species cannot be overemphasized. This is the goal of solid wood products research.

Mechanical Properties of Adhesives

Research on the mechanical properties of epoxy adhesives in free film form yielded useful quantitative data on the effects of elastomers on tensile strength. These additives were introduced to improve the elastic properties of the cured adhesive film, with the object of developing a bond that would respond to dimensional changes in wood. These data from free films provided valuable information on adhesive properties for comparison with their behavior when confined in butt joints in thick gluelines. Elastomers used were a polysulfide, a polyamide, a mercaptan-terminated acrylonitrile butadiene polymer, and an epoxy with a dimerized C-18 fatty-acid component. Each was blended with the epoxy resin in proportions of 5, 10, 20, and 40 parts per hundred parts of resin.

Tensile strength of free films of the elastomerized resin was lowered progressively in proportion to the quantity of elastomer added, declining overall from about 10,000 pounds per square inch to about 6,000. The significance of the decline is being evaluated in terms of the counter-balancing benefits gained in improvement of the film's elasticity.

Production Cure of Finger Joints

A rapid platen method of partially curing room-temperature-setting adhesives used in finger joints was found to have good possibilities, especially for small commercial operations for which high-frequency heating equipment and skilled operators might be too costly. At a platen temperature of 450° F., resorcinol-glued specimens attained tensile strengths of 1,100 to 1,200 pounds per square inch 10 minutes after a 40-second platen application. This strength is considered adequate for removing stock to a storage area for completion of the cure. Twenty-four hours later, similarly made joints broke in the wood rather than the adhesive, indicating substantially complete cure of the resin.

Resins for Gap-Filling Repairs

The seasoning checks that can develop in wood structural members can be filled with epoxy resin with beneficial effects on shear strength, experiments showed. Artificial checks were cut into sections of glued laminated beams and repaired with three epoxy resins found suitable from earlier tests. The beams were then put through soaking-drying cycles and exposed to high and low relative humidities, after which they were tested for shear

A railroad tie with a veneer "cap" glued on it to prevent surface checking is examined by Albert C. Higgins, FPL technician who conceived the idea.

14

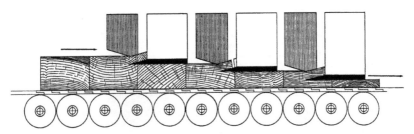

Drawing showing a multiple-knife arrangement for possible use in conjunction with the multiple-flitch concept for slicing wood up to 1 inch thick.

M 134 597

strength. The repaired beams were little different in stiffness and strength than unchecked beams.

Veneer Cap Prevents Checks

The deep checks that often occur in railroad ties, utility-pole crossarms, bridge stringers, and other wood timbers used outdoors can be prevented by gluing a veneer "cap" to the timber, experiments now under way indicate. The cap effectively restrains the development of these separations in wood when the veneer is glued on with the direction of its grain at right angles to that of the timber. The method is expected to provide an inexpensive solution to an age-old problem that frequently results in weakening of beams, loosening of fastenings, infection with decay fungi when the checks expose wood not reached by penetrating preservatives, and premature replacement. A patent has been applied for to assure free public use of the method.

Presurfacing Aids Fast Drying

Drastic reduction of drying costs was shown to be possible in experiments in which 4/4 red oak was dried to 5 percent moisture content in 14½ days. Drying time was shortened by 27 percent by using a kiln schedule of temperatures and relative humidity based on new knowledge of drying strains in northern red oak and the new technique of presurfacing the wood before seasoning it.

Indications from the experiment are that the commercial practice of air drying followed by kiln drying can also be greatly shortened. About 2 weeks of forced-air drying and 5 to 7 days of kiln drying appear to be sufficient to season the 4/4 red oak.

Presurfacing was also shown to be effective in preventing checking of green beech, one of the hardwoods most prone to check during kiln drying. Presurfaced beech squares checked only half as much as rough-sawn squares. Turning of the green squares to rounds reduced surface checking by about two-thirds and also significantly reduced bowing of the 3-foot specimens. Turning also, of course, substantially increased kiln capacity.

Multiple Flitch Slicing

A new system of producing sliced wood up to 1 inch thick at greatly increased production rates was experimented with on the Laboratory's special equipment. The system is envisioned as employing either a single knife to cut several flitches in a tandem arrangement, or a series of knives set at staggered heights to cut flitches moving past them continuously.

The multiple flitch arrangement appears to alleviate damage to the wood that occurs at both the start and the end of the cut when a single flitch is cut. The continuous cutting system increases production about 4 times as compared with a single knife cutter at a given cutting velocity.

Wood Flaker Also Planes

An experimental disk-mounted knife arrangement for flaking lumber was found to do an intermediate job of smoothing the wood while producing flakes worth twice as much for particleboard as the shavings of a conventional planer. The experimental device rotates cutting blades across the lumber from side to side of the piece, rather than lengthwise as a planer does. Knives can be step-mounted on the disk to do a continuous job of planing.

Measuring Smoke Production

The use of smoke production as an acceptability criterion for building materials by code officials has opened a new line of research at FPL, the object of which is to develop a realistic method of measuring smoke density. The presently accepted method is based on light-obscuration data obtained as part of the 25-foot tunnel test for surface flame spread adopted by the American Society for Testing and

Materials. The smoke development part of this test appears to have many technical faults and to lack the precision attributed to it by code authorities.

To develop a more meaningful method, FPL researchers have adopted a smoke chamber developed by the National Bureau of Standards. In this chamber, equal areas of test specimens are exposed in all tests, conditions of heating and flaming can be varied to obtain basic data, and the results can be related to full-size volumes, test areas, and light paths.

In preliminary experiments with this smoke chamber, data were obtained on four wood-base materials. Similar investigations are under way on nine wood species. Future work will include evaluations of wood panel products, paint systems, and chemically treated wood.

In fire-endurance studies a simple test was developed for comparing the fire resistance of different adhesives near fire-exposed surfaces, and a door with a medium-density particleboard core was shown to pass the standard 45-minute fire-resistance test for use in public buildings. A liquid ammonium phosphate fertilizer developed by the Tennessee Valley Authority was found to be an effective fire retardant, equivalent to di- or monoammonium phosphate, conventional fire retardants, in protection afforded wood from fire, corrosivity, hygroscopicity, and strength impairment. The fertilizer also is 30 to 40 percent cheaper.

Marine Treatments for Wood

Promising results with a dual pressure treatment of wood, first with a water-soluble preservative and then with creosote, led to a major study of preservative performance in sea water, in cooperation with the Naval Facilities Engineering Command. Experimental treatments had shown that the dual treatment, made in accordance with a specification of the American Wood-Preservers' Association, achieves sufficiently high retentions of preservative in southern pine to warrant strong hope of long service life in waters infested with marine borers that attack wood. Douglas-fir, generally more difficult to treat adequately, is also being evaluated for treatability by the dual process.

Post-size specimens were installed under a dock at the Key West, Fla., Naval Station in November 1968. Some of these had been treated with a selected water-borne preservative salt, some with creosote, and some with both by the dual process. Half of the wood specimens had also been infected with mold fungi before treatment to increase their porosity and permit deeper penetration. The ability of the mold-infected wood specimens to retain preservative under such conditions for reasonably long service life is in doubt, and will be closely watched. Changes in the amounts and quality of the preservative

16

In shipworm-infested sea waters at Key West, Fla., the efficacy of preservative treatment for wood piling that had been exposed to mold fungi to improve preservative penetration is being compared with like treatment of piling not exposed to molds. Here Technologist Lee Gjovik lowers a pile specimen into the water.

Chemist Henry G. Roth takes boring from treated wood for analysis of preservative content, in research to develop results-type specifications for preservative treatment. With appropriate specifications, the user of treated wood could determine for himself the quality of treatment by assaying borings from representative treated items.　M 133 784-2

chemicals will be studied with improved analytical apparatus and techniques.

Preservative Treatment of Lumber

A simple vacuum treatment of construction lumber holds promise as adequate for a number of species to give good preservative protection under moderate weathering conditions, experiments indicated. The experiments involved 2-inch dimension of Douglas-fir, white fir, western hemlock, lodgepole pine, ponderosa pine, and southern pine.

The treatment consisted of drawing air from the seasoned lumber by means of vacuum for about 30 minutes in a cylinder, then flooding the cylinder with acid copper chromate, a water-borne preservative, or oil-borne pentachlorophenol. The lumber remained submerged in the preservative for 7½ hours at atmospheric pressure.

With the exception of ponderosa pine heartwood, all species had retentions of acid copper chromate sufficient to meet requirements for the intended use conditions. Retentions of pentachlorophenol in white fir, incised western hemlock, and southern pine were adequate to meet specifications for pressure-treated lumber in fresh water as well as in the ground or above ground. Preservative penetration of the sapwood of the three pines was practically complete, but heartwood was not nearly so well penetrated as by a pressure treatment.

Termite Suppression

For the second year, a termite attractant-insecticide was found to be effective in suppressing foraging of the wood-destroying insects on small isolated plots in Ontario, Canada. The experiments with the attractant, developed at FPL, are being conducted in cooperation with the Canadian Forest Products Laboratory.

Second-year inspections in July 1968 were favorable enough to raise the possibility that eradication rather than temporary suppression may be occurring. The attractant-insecticide material was set out in small plots and around buildings. A decayed-wood attractant lures the termites, which burrow through the ground in search of food. The exact mode of suppression is not yet known. But in plots where insecticide-treated decayed wood is set out, termite activity ceases.

Additional observations were initiated last year to ascertain the nature of the changes taking place in the termite colonies. These, it is hoped, will provide definite evidence as to whether termites are being eradicated from the treated plots.

The same poisoned bait is being tried in new experiments in the South, where subterranean termites are much more prevalent and plentiful. These experiments, being conducted in cooperation with the Southern Forest Experiment Station, New Orleans, La., will provide evidence of the suppressive capability of the treatment under those conditions. Included in those experiments are the destructive Formosan termites that have recently appeared along the Gulf Coast.

Meanwhile, a study of the life cycle and habits of subterranean termites is yielding important basic information. In Wisconsin, colonies of the species *Reticulitermes flavipes* have been shown to survive the cold winters. The spread of the insect, however, is very slow in most infestations, and mainly by underground travel rather than by the swarming of flying insects that occurs in warmer climates.

An accelerated laboratory method of determining the termite resistance of different species of wood and the protection afforded by different preservatives was developed to a stage that should make it acceptable for tentative standardization. The termite hazard imposed by the method seems to be at least as great as ordinary field exposures. Evaluations of resistance are extremely simple, requiring only measurement of weight loss in the wood blocks under test.

WOOD ENGINEERING RESEARCH

FPL research on the mechanical properties of wood and on structural design of wood and wood products made significant progress in both heavy and light engineered construction and in packaging. Among developments were:

1. Evaluation of some of the largest laminated wood beams ever studied contributed knowledge of how to design them more safely.

2. Design of structural members with hardboard, utilizing criteria developed in earlier FPL research, demonstrated the validity and practical utility of the criteria.

3. Safer, more shock- and vibration-resistant packaging was made possible by experiments simulating transit conditions by rail and truck.

4. Adhesive-bonded pallets survived use tests in truck shipment of beer in better condition than nailed pallets.

Strength of Glued-Laminated Beams

Research previously reported on the strength properties of some of the largest glued-laminated wood beams ever tested—50 feet long, 31½ inches deep, and 9 inches wide—revealed that additional design criteria were needed for such large members. Specifically, results showed the need for upgraded tensile laminations.

Based on this work, industry developed and published specifications for upgraded tensile laminations for structural glued laminated members. Requirements were adopted for the use of these quality laminations for all large beams in order to meet AITC certification requirements. To check the validity of this increased quality tension lamination specification, 26 beams were subsequently manufactured and tested at the Forest Products Laboratory; six were 50 feet long and twenty were 40 feet long. One-half were of Douglas-fir and the other half of southern pine.

Although results of these experiments are still under evaluation, cursory analysis of the data indicates that the strength of large glued-laminated members is very dependent upon the strength of the tensile laminations and of the end joints in the laminations. Knots with associated localized cross grain and finger joints were the primary triggering agents in beam failures. The findings clearly indicate that knowledge of the strength and performance characteristics of finger joints in tension is deficient and further research is needed.

Beam-Deck Stability

More economical structural systems employing glued-laminated beams to support deck-type roofs are expected to result from new design data developed from research on the elastic stability of several commercial constructions. The work was done in response to needs of laminators and building de-

This 50-foot-long beam, 31½ inches deep, was one of a number tested to failure in experiments to determine the strength contributed by bottom tensile laminations of lumber higher in quality than that used in the other parts of the beam.
M 135 614

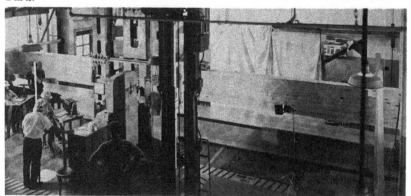

signers for data on narrower and deeper beams. Such beams are preferred because they are stronger and stiffer than wider and shallower beams of comparable weight.

The design problem arose from the fact that beams of high depth-to-width ratio lose lateral stability, hence require adequate lateral restraint to prevent them from buckling. The FPL analysis developed data on shear stiffness and other properties that demonstrated the contribution to lateral stability provided by the decking, which heretofore has not been recognized. Availability of the new data is expected to stimulate use of the more efficient deep beams.

Design Stresses for Wood Piling

Research is under way at FPL and elsewhere to establish badly needed design stresses for wood piling used as foundation supports for large buildings. Except in a few local building codes, such design values are nonexistent. Lack of a national standard is believed to be one reason why wood piles have been losing this valuable market to piling made of other materials.

Basic data on compressive and bending strength of three piling species, southern pine, Douglas-fir, and red oak, were obtained in a series of experiments conducted in cooperation with the American Wood Preservers' Institute. Fifteen 50-foot class B piles of each species were used.

Compressive strength of the Douglas-fir and red oak piles was shown to be high enough to make these species competitive with other materials. Southern pine piles were about 25 percent lower, but much of this deficit was attributed to the effect of steam conditioning before the piles had been preservative treated. It had been established from previous research on wood utility poles that such conditioning lowers strength much more than kiln drying. From recent data on pile sections, it was concluded that kiln drying lowered the strength of the southern pine piles by only about 12 percent.

Cooperation is continuing with the AWPI and the American Society for Testing and Materials to develop reliable timber pile design stresses.

Hardboard Structural Uses

The utilization of hardboard as an engineering material for structural purposes is a long step nearer reality with the completion of the FPL study of its mechanical properties. Only creep under long-continued loading remains to be established.

To demonstrate the applicability of the new data, an I-beam 34 feet long and 16 inches deep was built with hardboard webs and lumber flanges, and uniformly loaded to failure over a 30-foot span. The ultimate load predicted on the basis of the panel shear strength of the hardboard was 19,800 pounds

A 34-foot-long girder with an experimental web of hardboard and flanges of lumber was described by Wayne C. Lewis, FPL engineer, to four members of the Northern Illinois Hoo Hoo Club, all Rockford area lumbermen. From left are Bruce B. Harvey, manager of a Loves Park, Ill., lumber company; Virgil Harbolt, lumber salesman at Kirkland, Ill.; C. E. Sandstrom, and Wayne A. Oliver, both Rockford lumber distributors.　　M 134 305 F-3

per square inch. Average ultimate load of three specimens was 19,600 pounds. Similarly, the predicted deflection at a design roof load of 40 pounds per square foot was 1.56 inch; actual average deflection of the three specimens was 1.70 inch.

Such a beam could be used to support a roof at a saving in weight and material of about 40 percent compared with solid wood.

The hardboard industry has been given the results of the research, which was done in cooperation with the American Hardboard Association. Industry, however, has not yet decided whether to produce one or more stress-rated grades of hardboard for engineering uses or to assign working stresses to mill-run material. Meanwhile, similar engineering studies are being made of medium-density hardboards in cooperation with a manufacturer.

Wood Fastenings Research

The concept of a nail as a beam supported on an elastic foundation—wood—has been partly verified by experiment as a possible approach to describing

19

the lateral resistance of nailed joints by theoretical analysis. The characteristics of the elastic foundation determine deflections of the nail and, thus, stresses in the nail and the wood. Variations in nail size and type appear to be adequately described by theoretically determined parameters. Experiments are being continued to establish values of the elastic foundation modulus. Indications are that it varies between one-third and one-half the modulus of elasticity of softwoods, and is about equal to the elastic modulus of oak and similar hardwoods.

A technique was developed for subjecting a nailed joint to cyclic lateral loading as a means of determining the performance of fastenings under repeated loads. The equipment and method are expected to make possible the direct study of variables affecting performance of a variety of wood fastenings.

Spruce Pine Properties

Spruce pine (*Pinus glabra* Walt.) is one of the minor southern pine species, occurring in the Atlantic coastal plain from South Carolina to northern Florida and westward to Louisiana. As such, it has not been so closely examined in the research laboratories as the more widely used species. In recent years, however, particularly with the emergence of the southern pine plywood industry, its use for structural purposes has brought a need for better knowledge of its mechanical and related properties.

Specimen material for the standard tests was obtained by a new random sampling method described in the "Wood Quality Research" section of this report as used to collect samples of Engelmann spruce during the summer of 1968. The method, known as proportionate probability sampling, is based on volume and geographic distribution of a species.

The mechanical properties of spruce pine were found to be considerably below those of the four major southern pines, longleaf, slash, shortleaf, and loblolly. In these properties, the wood is quite comparable to white fir. Spruce pine in the green condition was found to have an average modulus of rupture of 6,000 pounds per square inch; modulus of elasticity of 1,000,000 pounds per square inch; compressive strength parallel to grain of 2,840 pounds per square inch; compressive strength perpendicular to grain at proportional limit 280 pounds per square inch; and shear strength parallel to grain 900 pounds per square inch.

Moisture Meter Charts

Correction charts were published for radio-frequency moisture meters of both the power-loss and capacitive admittance types. These charts provide correct moisture content values for wood based on meter readings made at various temperatures. Resistance-type meters were included in the study and these results confirmed previously published correction charts.

Container Shock, Vibration Resistance

Research on the effects of shock and vibrations imposed on shipping containers during transit has thrown new light on packaging improvements needed to reduce the heavy loss and damage claims — about $200 million annually — incurred during rail and truck shipment.

The research is producing new information on the effects of stacking several loaded containers one atop another in railroad cars and trucks. This practice has been shown to be much more damaging in moving vehicles than in a warehouse. During transit, the loads imposed on the bottom container in a stack may increase greatly because of vibrational excitation of containers and their contents.

Research on verticle dynamic loading of a corrugated fiberboard container was conducted in cooperation with the Fibre Box Association. The first phase of this research consisted of a laboratory study of the effects on an empty container of a stacked load coupled with vibrational forces simulating transit conditions.

For test purposes, a simple loading configuration, which can be analyzed as a single spring-mass system, was used in which the stacked load represents the mass and the container the spring. Vertical vibrations were imposed on the bottom of the container to simulate those exerted by a railroad-car or truck floor during movement. The effects of different frequencies of vibration which exist in rail and truck transportation were studied in relation to the natural frequencies of the load-container systems.

From basic vibration theory, it is known that the magnitude of the forces transmitted to the load is a variable ratio of the applied vibrational forces. This ratio, called the transmissibility ratio, is a function of frequency. When the frequency of the applied vibration approaches the natural frequency of the load-container system, the vibrational forces transmitted to the load are magnified and may become much greater than the applied vibrational level. This condition, called resonance, subjects the containers to high dynamic loading forces.

The spring-mass system approach to the study of vibrational forces exerted upon containers in rail and truck transit is greatly simplified as compared with actual transit conditions. It is regarded as an essential first step, however, in any study of those conditions which has as its objectives a better understanding of the factors involved in container damage during transit and the development of safe dynamic loading limits for containers. Experiments have confirmed the relationship between safe load-

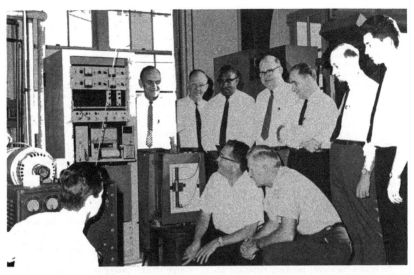

Members of a Department of Defense Packaging Advisory Group observed new packaging research techniques June 5 at FPL. Shown watching an experiment in vibration effects on container contents by W. D. Godshall, FPL electronics expert, are (standing, from left) William Kurtz of Defense Supply Agency Headquarters packaging staff, Alexandria, Va.; C. Y. Best, director, Joint Military Packaging Training Center, Aberdeen, Md.; R. L. Duckett, packaging specialist, Marine Corps Headquarters, Washington, D.C.; N. J. DeMars, director of packaging, Tobyhanna, Pa., Army Depot; R. C. Young, packaging specialist, U.S. Naval Supply Systems Command, Washington, D.C.; James R. Green, packaging specialist, Headquarters, Defense Supply Agency Contract Administration Service, Alexandria, Va.; and Peter Brunner, packaging specialist, Air Force Systems Command, Washington, D.C. In foreground are Frank G. Harris, packaging specialist, Air Force Logistics Command, Wright-Patterson Air Force Base, and Col. V. C. Ramberg, training chief, Army Materiel Command, Washington, D.C.

M 134 891-2

ing limits and the vibration frequency of the dynamic loads applied to the container. It has been clearly demonstrated that loaded containers are frequency-sensitive systems and that destructively high dynamic loadings can be developed when the frequencies of the applied forces approach the resonant frequency of the load-container system.

The shock-absorbing capacity of a package is concomitantly under study from the standpoint of the so-called "container effect" on cushioning efficiency in protecting packaged goods from shock forces. This approach examines the overall capacity of container and interior cushioning to absorb shocks and recognizes that under certain conditions the container may actually lessen the protection afforded by the cushioning.

In exploratory experiments, dummy loads of various weights were cushioned in two kinds of containers, cleated plywood and corrugated fiberboard.

Instrumentation within the load measured the acceleration, in gravity units, generated when the package was dropped 2 feet onto a concrete base. The cleated plywood boxes showed to clear advantage in helping absorb shock forces under the test conditions used.

The tests clearly showed that the container effect must be considered if both overpacking and underpacking are to be avoided. For example, a 4-pound load supported by a 25-square-inch cushion in a cleated plywood box experienced a peak force 20 percent greater than was predicted on the basis of the protection afforded by the cushion only; on the other hand, a 10.5-pound load in the same box and with the same cushion experienced a peak force one-third less than predicted.

These experiments demonstrated that the shock spectrum, graphically represented as the natural vibration frequency of hypothetical mechanical elements of the load plotted against their peak response acceleration, is a more accurate criterion of damage potential than is the peak acceleration of the load in a cushioned package.

A study of the cushioning properties of corrugated fiberboard pads is producing design criteria needed to permit use of this material in packaging. Data on the flat-crush cushioning capability of such pads were published.

Adhesive-Bonded Pallets

The performance of pallets assembled with a mastic adhesive and matching ones conventionally nailed together is being demonstrated both in laboratory long-term tests and actual industrial performance.

The comparisons were undertaken to explore the potential of adhesive bonding from a service standpoint. Information on this is needed to determine whether adhesive bonding would be worthwhile if it should become economical from a production standpoint—a condition not yet established. Theoretically, adhesive bonding is intriguing because nailed pallets made of green lumber—as is usually the case—are weakened as the wood dries, since the nails loosen and the deck boards split at the nails.

To explore the service potential, nailed and adhesive-bonded pallets were made and submitted to laboratory tests and service conditions. The adhesive used was a mastic that remains somewhat elastic indefinitely.

One nailed and one adhesive-bonded pallet sustained 2,000 falls in a rotary drum without damage immediately after they were made. Another pair were given a standard drop-corner test. The adhesive-bonded pallet took the 6 free-fall drops from a 40-inch height without deformation, but the nailed one was deformed at least 1⅛ inches along the diagonals.

After being stored outdoors for a year, the pallets were retested. Again the adhesive-bonded pallet survived in the revolving drum without failure, although a deckboard fell off after 1,800 falls. The nailed pallet lost three deckboards after only 560 falls and was considered to have failed.

The corner-drop retest of the other two pallets gave the same results as the first test: No distortion of the adhesive-bonded pallet but a total distortion of 1½ inches in the nailed one.

The pallets put into industrial use were inspected after 1 year of transporting case beer by truck from a Milwaukee brewery. The adhesive-bonded pallets held up extremely well, with no records of any failures. The nailed pallets remained serviceable, but inspection revealed some splitting of deckboards, especially around the nails—as would be expected from wood shrinkage. A workman in charge of loading the pallets in an automatic stacker reported that the adhesive-bonded pallets performed exceptionally well in the stacker because they remained true and square.

Corrugated Fiberboard Patent

A machine that makes corrugated fiberboard with the wood fibers in both the corrugating medium and the liner sheets alined parallel to the flute direction—that is, vertically in the container walls—was patented by Keith Q. Kellicutt, engineer in charge of FPL packaging research. Conventional machines produce fiberboard with the fibers oriented crosswise to the flutes. The orientation developed by Kellicutt's invention results in boxes with 30 to 40 percent greater stacking strength. The invention is covered by United States patent No. 3,425,888.

Paperboard Folding Patent

A patent covering a process of flexing horizontal scorelines of resin-treated fiberboard scored for container formation was granted an FPL packaging technologist, John W. Koning. The process facilitates box formation without breaking or cracking the resin-embrittled fiberboard along the score lines. It is covered by United States patent No. 3,374,715.

WOOD CHEMISTRY RESEARCH

FPL's wood chemistry research is oriented toward the mission of finding new ways to utilize wood, improving existing ways, and obtaining the knowledge of its chemical properties needed to overcome various deficiencies that complicate processing, incur waste, and limit usefulness of wood products. Attainment of these objectives involves a three-pronged research attack divided about equally into investigations of whole wood's durability and characterization of its chemical components that impart properties desirable and undesirable; delignification and carbohydrate chemistry; and investigations related to pollution abatement, such as the chemical deterioration of wood and the utilization of harvesting, milling, and other residues.

A substantial proportion of research in this area during 1968 was devoted to assaying the potential utility of wood as roughage and feed for livestock (see 1967 Annual Report). Other research advances included the development of analytical techniques for identifying the constituents of tall oil, a pulp mill byproduct; exploration of dehydration reactions of glucose under acidic and alkaline conditions; a procedure for analyzing the hydrolysis products of hardwood hemicellulose; x-ray diffraction studies of cellulose; and completion of a new nomenclature for diterpenes.

Wood as Livestock Roughage and Feed

Experiments designed to demonstrate the value of wood as roughage for livestock, especially ruminants (cud chewers), continued to yield favorable results. In cooperative feeding experiments at Auburn and Penn State Universities, cattle and sheep fed red oak sawdust in concentrations up to 30 percent showed no adverse effects as to weight gains. The roughage in these experiments was studied for its value as a stimulant for saliva production to neutralize acidity produced in the rumen of the animal by bacterial action on feed concentrates. Red oak was used in these experiments because it is the species of primary interest to the Tennessee Valley Authority, which is supporting the investigation.

The broad implications of the work are apparent in statistics which show that, of a total cattle population of 100 million head in the United States, some 15 million are regularly fed highly concentrated diets for rapid weight gains in feedlots near large centers of population. Shortages of hay in these

Chemist Richard C. Weatherwax operates apparatus built by FPL scientists to measure surface area of cellulose, wood, and wood pulps. It is based on the Brunauer-Emmett-Teller method for determining surface area. The specimen is placed in a vacuum, and nitrogen gas collects on it, the amount being a measure of the surface area.

23

areas prompt feeders to look for other roughages. Even a minor fraction of the total feed requirements for these cattle would utilize substantial amounts of wood as roughage. For example, if 10 percent of the feed consisted of wood, it is estimated that some 3 million tons would be needed.

Meanwhile, other experiments are being carried out to study the potential of wood not only as roughage but as a nutritive feed that livestock can convert to energy. In the form of sawdust, no wood species approaches the feed value of hay, although some are much more utilizable than others. Among hardwoods investigated in laboratory experiments and actual feeding studies, aspen was shown to be most digestible and the oaks least—virtually worthless. But even aspen rated well below alfalfa hay in digestibility. A major factor interfering with digestibility appears to be the lignin content of wood.

Experiments to improve the digestibility of wood are aimed at making the carbohydrate more readily accessible, either by alteration of the lignin-carbohydrate complex or by grinding the wood to such small particles that digestion can occur. Chemical pulping processes that remove lignin are generally too costly. Experiments were therefore conducted with inexpensive sodium hydroxide which causes extreme swelling through breakage of ester crosslinks. Aspen treated with a 1 percent solution approached the digestibility of alfalfa in laboratory experiments utilizing the stomach juices of cattle to compare digestibility (see 1967 Annual Report). The digestibility of aspen rose from 35 to 52 percent. The next best gain was by silver maple, which increased from 18 to 41 percent. Red oak increased from 3 to 14 percent—still far below the 50 to 60 percent digestibility of alfalfa.

By limiting the volume of aqueous solutions of sodium hydroxide (at ½ or 1 percent concentrations), enough information was gained to make possible the calculation of the theoretical amount of alkali necessary to derive the maximum benefit from this simple treatment. Although the theoretical alkali requirement (based on the wood) is 6 percent, even if a 10 percent requirement is assumed, $6 worth of sodium hydroxide costing 3 cents a pound would be needed to treat a ton of aspen. The digestibility of the aspen would be in the range of 50 to 55 percent.

The most successful treatment of all on a laboratory scale was vibratory grinding to extreme fineness. Digestibility of aspen and oak both increased to about 60 percent. Grinding in commercial equipment designed for taconite grinding, however, resulted in only a small improvement, thus casting serious doubt on the commercial value of the finding.

Tall Oil Analysis

Analytical techniques were originated that are expected to prove highly useful to industrial chemists in the utilization of tall oil, a byproduct of pulp mills. Various useful chemicals are obtained, but the chemical reactions involved have not been well understood. Fatty acids and resin acids are obtained from the black spent liquor. The fatty acids in particular have risen in value in recent years. Tall oil rosin also is now of very high quality, comparing favorably with gum rosin.

Gas chromatography techniques were developed for analysis of tall oil constituents. To facilitate identification of these constituents, spectra were obtained by ultraviolet, infrared, and nuclear magnetic resonance spectrometry; these spectra are now available and will be published. The work was done in cooperation with the Pulp Chemicals Association. It is continuing with analytical studies of the resins obtained from five southern pines.

Glucose Dehydration

New understanding was gained of the reaction products formed and the mechanism of their formation in glucose dehydration experiments with deuterium (from heavy water) as a tracer. The experiments were carried out in both acid and alkaline media to study those reactions occurring in the production of furfural and other furan derivatives and those reactions responsible for the loss of carbohydrate in conventional pulping procedures.

On the basis of reported mechanisms, the presence of deuterium in the reaction products (hydroxy methyl furfural in acid medium and metasaccharrinic acid in basic medium) was predicted. Using nuclear magnetic resonance procedures, deuterium was shown to be absent in the former and present in the latter. These results are important for understanding the role of 3-deoxyosone as an intermediate in the acid- or base-catalyzed degradation of glucose. This addition to the knowledge of these chemical reactions helps to explain why, for example, conventional pulp yields do not approach those of processes using polysulfide and borohydride.

Analysis of Hemicellulose

Work was completed on a thin-layer chromatographic procedure for analysis of hardwood hemicellulose. The procedure is being used to resolve disparate views outlined in the literature regarding xylan composition.

Cause of Odors Traced

Gas chromatography was used to run down the cause of offensive odors emanating from plywood made of virola, a Latin American species. The source was traced to certain lower fatty acids found in the wood. The acids are believed to be formed by bacterial action on starch granules in the parenchyma cells while the logs are being ponded or floated to loading docks.

Crystallinity of Cellulose

X-ray diffraction procedures were used to study the crystallinity of cellulose as an index of its condition in lignocellulosic materials. The degree of crystallinity, the size of individual crystallites, and their orientation are all believed to influence certain structural properties of wood.

X-ray crystallography is also being used to study changes in crystallinity of pulps subjected to different mechanical procedures. Fine grinding in a vibratory mill is shown to destroy the crystallinity of dry cellulose. Beating of pulp, however, takes place in water, and exposure to water or high relative humidity is shown to restore a high degree of the original or modified crystallinity in the cellulose. It is planned to make x-ray diffraction studies of celluloses beaten to various freenesses to learn the extent of changes in the quality and quantity of crystallinity.

Cellulose grinding, pulp beating, and similar processing techniques also increase the specific surface area of the material, and this factor is also be-

24

ing studied for its possible effects on pulp fiber bonding and digestibility by micro-organisms. A device for determining surface area of material, using the Brunauer-Emmett-Teller analysis, was constructed. It determines surface area of a material by measuring the amount of nitrogen gas that collects on the surfaces.

Nomenclature of Diterpenes

An FPL-devised system of nomenclature for diterpenes that greatly simplifies the existing system was submitted to the International Union of Applied and Pure Chemistry for approval. An international committee of 11 prominent chemists was named to bring it to final form for adoption. The new system will appear as a chapter in a forthcoming book on these nonpolar compounds, which have been intensively studied at FPL as extractives of wood and bark.

diffraction pattern of highly ordered cellu-
tio of the intensity of discrete diffraction
perimeter to that of the general amorphous
a measure of the degree of crystallinity of

WOOD FIBER PRODUCTS RESEARCH

Aspen, the once-despised hardwood that invaded the Lake States after the virgin pine forests were removed, may become the raw material for a newsprint mill in northern Minnesota as a result of FPL research.

Successful experiments confirmed that high-quality newsprint can be made from groundwood pulp of this species, which is plentiful in the area, and as a result plans were formulated by Minnesotans to build a pulp and paper mill. Because the mill will bring industry and employment to an economically depressed area, a loan was sought from the Economic Development Administration.

Aspen had earlier found considerable use as a raw material for chemical pulps, starting with the development at FPL of the neutral sulfite semichemical process in the 1930's. Aspen semichemical pulp found its first and greatest use as a corrugating medium for containerboard.

Groundwood pulp offers greater economies in production, however, because virtually all of the wood is utilized. In chemical pulping, utilization ranges from less than one-half to about three-fourths of the wood substance, depending on the process used. Polluting byproducts are also avoided. Only a small proportion, about 20 to 25 percent, of chemical pulp is used in newsprint to add strength.

The newsprint experiments were begun when several organizations indicated interest in promoting the development of a newsprint mill utilizing aspen. Among these organizations were the Northern Great Lakes Resource Development Commission, the Wadena County Resource Development Commission, the Otter Tail-Wadena Community Action Council, and the Minnesota Department of Conservation. Research findings promoted the organization of the Minnesota National Paper Co. to obtain financing and construct and operate a mill. To begin with, the proposed mill would have a capacity of 360 tons of newsprint a day. Chemical pulp needed would be brought from other sources.

Newsprint comparing favorably in most properties with commercial newsprint was produced from aspen groundwood pulp. Among properties evaluated were bursting, tearing, and tensile strength, brightness, opacity, printability, and strike-in—the last a measure of ink show-through on the back of the sheet.

A number of newsprint mills formerly operated in the Lake States, utilizing spruce, balsam fir, and

FPL's corrugator was used in an investigation of the runnability of various kinds of corrugating medium furnished by manufacturers in the United States, Canada, and Europe. Higher corrugating machine speeds are the goal.
M 136 064-4

The cottonwood trees in this experimental Forest Service plantation at Stoneville, Miss., were 4 years old when this photo was taken. FPL experiments show pulp made from such fast-grown trees does not differ importantly from pulp of normal cottonwood. (Photo courtesy Southern Forest Experiment Station.)

other softwoods. Economic factors, however, led to their discontinuance or shifts to higher-value products, such as tissues and toweling, for which northern softwoods are especially suitable. Some 75 percent of United States newsprint needs is now imported from Canada and other countries. Total consumption is some 9 million tons annually.

Cottonwood for Paper

Exceptionally fast-growing cottonwood trees grown on a Forest Service plantation near Stoneville, Miss., from cuttings of selected parents were pulped by the kraft process in experiments to evaluate their suitability for paper.

Their papermaking properties were found to be comparable to those of normal cottonwood. The trees had been grown at a plantation of the Southern Hardwood Laboratory, maintained by the Southern Forest Experiment Station, New Orleans. Trees estimated to be 7 years old had attained heights up to 40 feet. Bolts shipped to FPL ranged up to 10 inches in diameter. The four-foot bolts were taken from various sections of the tree trunks. Average growth rate was 2.85 rings per inch of diameter.

Magnesium-Base Semichem

Substituting magnesium for soda in neutral sulfite semichemical pulping liquor was found to be practical in experiments in which the pulp chips were first impregnated with small quantities of ammonium hydroxide or bisulfite to permit better penetration of the pulping liquor. The advantage of magnesium over soda is that it is reusable when recovered, whereas soda is not, being recovered as sodium sulfate for shipment to kraft mills at no substantial economic benefit.

The magnesium-base semichemical liquor was used to pulp white oak, which is particularly difficult to penetrate because hairlike growths called tyloses plug its vessels. Preimpregnation with 2 percent of ammonium hydroxide or bisulfite permitted adequate penetration of the pulping liquor.

Polysulfide Pulping

The mechanism whereby socalled end-group peeling is prevented by pulping liquor in the polysulfide process was defined in a study concluding work on this promising high-yield modification of the conventional kraft process.

Experiments showed that gluconic acid is formed by oxidation of the end groups of a typical glucose chain in polysulfide pulping. The gluconic acid prevents loss of substance that would otherwise occur from chemical reactions that cause progressive weakening of the bonds between the glucose units in the chain and socalled peeling of loosened units. Losses can run as high as 10 to 15 percent by weight. Gluconic acid is actually formed by epimerization of mannonic acid, which occurs first but is transformed into the closely similar gluconic acid in a continuation of the polysulfide reaction.

Research having shown the feasibility of the polysulfide process for raising yields of kraft-type pulps as much as 20 percent, the process, complete with a chemical recovery system, is now available to industry for pilot-scale trials and commercial use.

Stabilized Hardboard

The dimensional stability of hardboard under widely fluctuating conditions of relative humidity was shown to be markedly improved by applying small amounts of phenolic resin to the surface just before final compression takes place in manufacture. Boards so treated withstood repeated cycles of high and low moisture conditions much better than untreated control panels. Bending strength and abrasion resistance were also better, and appearance was improved.

Experiments are being continued to compare the effectiveness of the treatment on hardboards made of hardwoods and softwoods and the influence of such variables as fiber preparation, resin application, and press temperature. With the advent of structural hardboards engineered for load-bearing uses, their dimensional stability takes on added importance.

Fiber Bonding in Paper

A new method of measuring by electrical conductivity the hydrogen bonds that hold pulp fibers together in the sheet yielded significant new information on the strength of papers made from either springwood or summerwood fibers of southern pine.

Using the new method, conductivity values were

Bernard Toker, right, chemical engineer at the Israel Fiber and Forest Products Research Institute in Jerusalem, spent 3 months at FPL on a United Nations fellowship to study scientific developments. Here, with Vance Setterholm, FPL technologist, he examines equipment used to evaluate strength of paper.   M 135 571

established for individual pulp fibers. Then pulp handsheets were made from those fibers, and the conductivity of the sheets was measured by the same method. The difference between the conductivity of the fibers and that of the sheets was taken as a measure of the hydrogen bonds formed between fibers in the sheet.

Measurements of southern pine handsheets in this manner revealed that springwood fibers, being more flexible, develop more interfiber bonds and, therefore, stronger test sheets than the inherently stronger summerwood fibers. This finding suggests that the growing of southern pine with high summerwood content, and therefore high density, will not necessarily produce pulpwood that can be converted to paper of highest strength.

In another approach to determination of fiber bonding forces, the shrinkage forces generated in paper being formed on the paper machine were measured. Forces as high as 500 pounds per square inch along the web edge were measured in papers made from well-beaten furnishes. Taken as an indicator of the bonding potential of a pulp furnish, such data are believed to be of value in the design of papers and the operation of paper machines.

WOOD QUALITY RESEARCH

FPL research in wood quality deals directly with the quality of the Nation's total forest resource. The factors that affect quality, and the trends in wood quality that occur as the character of the resource changes, are significant areas for investigation. The information obtained often has immediate practical value to the wood-using industries and to consumers of wood products. Long-range benefits will accrue to the Nation through the growing of higher value forests for the future.

Among research highlights of 1968 were investigations into new ways of automating sawmills for more efficient production of lumber; evaluation of a more efficient sawing method for lodgepole pine; use of a new statistical sampling method to obtain a more valid sample of Engelmann spruce throughout its high-elevation range in the Rockies; studies of the effects of nutrition on wood structure; a study of the causes of color differences in black walnut; and new knowledge of the causes of stains and rancid odors in oak trees.

Automated Sawmilling Decisions

Electronic signals created by pulses of sound that can stab through a 1-inch board in thirty millionths of a second are being utilized in experiments aimed at automating the decisions involved in sawing hardwood logs into lumber. The work has as its immediate objectives the development of a technology and instrumentation that can sense and locate defects in hardwood boards and flitches, and a computer program that can control the edging and trimming decisions necessary to cut flitches into boards of lengths and widths that contain the greatest value of quality lumber possible. Longer term goals are the technological groundwork for computer control of tree bucking and log breakdown as well as flitch upgrading.

The system is based on a defect-detection device consisting of a pair of ultrasonic transducers, one on either side of the flitch. Pulses of sound pass from one to the other, and the elapsed time in microseconds is recorded electronically. Variations in elapsed time as a measure of sound velocity, and sound attenuation as a measure of pulse amplitude, indicate presence of defects.

The ultrasonic pulse is transmitted longitudinally along the grain of wood approximately three times faster than transversely or radially. This characteristic makes it possible to program a computer to locate knots and other defects closely associated with grain direction. Defects are located on the surface of the flitch by a simple x, y coordinate system, the x coordinate corresponding to the length of the flitch and the y coordinate to the width. A computer program has been developed to produce full-scale printouts locating defects for visual comparison with the flitches scanned.

Scanning is done under water for best results. Of nearly 10,000 transducer pulses recorded in evaluation studies with oak boards, only four were judged to indicate a defect in an area where none was visible. It was not determined whether a defect not visible on either surface might in fact be present within the piece.

The potential value of automating the actual decision-making involved in log breakdown and cutting up of flitches is great. The value of hardwood lumber produced in 1966 is estimated to be $650 million at the mills. Research has shown that, by proper edging, ripping, trimming, and crosscutting, it should be possible to raise the value of the lumber by $10 to $50 per thousand board feet, depending on initial grade and species. Overall, the value of production at the mill could be increased an estimated 13 percent by computerized controls. Thus the output of 1966 could have been worth $730 million. The cost of controls that automate defect sensing and decision making at the sawmill would be only a small fraction of the added value gained.

The research is confined to hardwood lumber for furniture, millwork, and other high-value products because the greatest potential rewards are here. Manufacturers of such products generally want only the clear material, called cuttings, that remains after gross defects such as knots are cut out. Some 7 to 8 billion board feet of oak, maple, walnut, and other hardwoods are used in the United States annually.

Hardwood Cutting Yields

FPL-developed charts from which yields of hard maple cuttings can be read directly according to grade of lumber and cutting size (see 1967 Annual Report) were found to be suitable also for No. 1 and No. 2 Common red oak lumber. In a check run at a Lebanon, Ky., mill, predicted yields for the No. 2 Common were within 1 percent of actual yield. On No. 1 Common, the predicted yield was about 4 percent under the actual yield, but examination showed that the lumber used was better than the average for the grade, thus accounting for the discrepancy.

Work was also begun to correlate the system with clear-one-face cuttings from black walnut lumber in the grades of Firsts and Seconds, First-One-Face, and No. 1 and No. 2 Common.

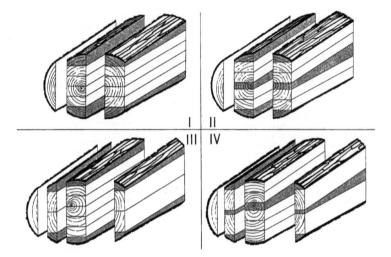

Methods used to saw lodgepole pine logs. Method I, conventional; II, FPL modified conventional; III, Scragg; IV, FPL improved Scragg.

Lodgepole Pine Sawing

Techniques for sawing lodgepole pine logs into 2-by-4 studs with least risk of warp were worked out on 680 logs that yielded 3,349 studs. The lumber was kiln dried and planed, after which all studs were measured for crook, bow, twist, and length, and the actual moisture content of each was determined. The basis of evaluation was the Western Wood Products Association's grading rules for studs.

Two commercial sawing methods, the conventional and the Scragg, were used to saw the logs, along with an FPL experimental method derived from the Scragg method. The Scragg and the FPL method were both markedly superior to the conventional method in reducing warp. The FPL method was better than either of the others in controlling crook.

Butt logs yielded studs more prone to crook and bow but less likely to twist than studs from upper logs. Logs dogged on the mill carriage so that their greatest radius was vertical yielded studs less inclined to warp and more likely to grade out No. 1. Studs from the larger logs were less likely to twist, but crooked and bowed about as much as the others. Studs from the inner area of the log near or including the pith showed significantly higher bow and crook than those from near the bark.

Engelmann Spruce Sampling

With helicopters, four-wheel-drive trucks, pack-horses — and frequently on foot — FPL scientists hunted high elevations of the Rocky Mountains from New Mexico to Idaho during the spring and summer of 1968. Their quarry was 50 Engelmann spruce trees from locations that had been selected beforehand by means of a new system of direct random sampling. Basically the system involves the selection of sample locations, trees within locations, bolts within trees, and test specimens within bolts, with probability proportional to volume at each stage.

The field work involved procurement of one 5-foot bolt from each tree. States from which trees were obtained (virtually the entire commercial United States range of the species), and the number of trees taken in each, were: Arizona, 1; New Mexico, 2; Colorado, 19; Idaho, 4; Montana, 10; Utah, 4; Wyoming, 6; Washington, 1; and Oregon, 3.

Location points were randomly selected in each state by latitude and longitude. Supervisors of National Forests where each location occurred were queried in advance as to whether Engelmann spruce grew within 2 miles of the location.

Locations often were in remote areas as much as 10 miles from any road. From those near roads, the

RANDOM SAMPLING...

took **FPL scientists into the Rockies for 50 Engelmann spruce trees afoot, on horseback, and in helicopters.**

1. A helicopter lifted one log from a bog-surrounded spot in the Boise National Forest of Idaho.

2. FPL technologists Robert Koeppen and Robert Maeglin muscled one log out of the Teton National Forest in Wyoming.

3. D. E. McDaniel, Meteetse, Wyo., rancher, used a packhorse to carry a log out of the Shoshone National Forest.

4. Colorado rancher Robert Stovall resorted to an old Plains Indian device, the travois, to haul this log out of the San Juan National Forest.

$$\frac{2 \mid 3}{1 \mid 4}$$

31

bolts were pulled out by men on foot, using a device invented by an FPL technologist that is driven into both ends and permits the log to turn when pulled. For longer distances, packhorses were used. In two instances, a helicopter was needed.

Bolts were taken to the nearest commercial shipping point for trucking to FPL. There they were cut into samples for evaluation of strength and related properties by the Division of Wood Engineering Research.

The collection task was by far the most far-ranging ever undertaken for a single United States species. It was not, however, the first FPL use of the sampling procedure. In 1967, spruce pine had been sampled in the same way. The procedure has come into use as the most reliable statistical sampling system yet devised.

Results of the evaluation will be compared with earlier data on Engelmann spruce. Existing FPL strength values are based on 18 trees from Colorado, Montana, and Idaho, mostly collected about 50 years ago. One of the objects of the work is to determine why the earlier samplings yielded strength values considerably lower than those established by Canadian authorities for the same species grown in Canada.

Wood Density Surveys

With completion of increment core collections in South Carolina, the density survey of the four major southern pines throughout the South was almost finished. Data obtained are being related to already published information on density elsewhere in the growth region of longleaf, shortleaf, loblolly, and slash pine, which extends from Texas to Virginia.

The information has proved invaluable to timber growers, forest managers, and processors of southern pine into lumber, paper, plywood, and other products, because specific gravity is a good index not only of various strength and related properties but of fiber content for papermaking. It was a key factor in the launching of the now booming southern pine plywood industry.

Meanwhile, a similar density survey is under way in Wisconsin, in cooperation with the North Central Forest Experiment Station and the Wisconsin Department of Natural Resources. Here for the first time hardwoods as well as softwoods are being surveyed for density. Field collection of increment cores is expected to be completed by mid-1969.

Through 1968, more than 5,500 cores had been received at FPL for specific gravity determinations and processing. FPL field crews also gathered specimens of the wood of trees of various sizes at different heights in the tree. From these specimens equations are developed that make it possible to estimate closely the specific gravity of the wood in a tree from a single increment core taken at breast height.

Plantlets from Aspen Tissue

Complete plantlets with a continuous root and shoot system were produced from aspen callus (isolated from living cambium tissue)—the first such growths ever achieved from woody tissue, although it has been done on herbaceous plants. The work represented a culmination of experiments in which either roots or stem shoots had been produced (see 1967 Annual Report).

Plantlet initiated from undifferentiated aspen tissue grown in vitro.

The plantlets were produced by excising shoots formed on aspen tissue and subculturing them on a synthetic medium lacking all growth hormones but retaining all the other nutrient ingredients, such as inorganics, vitamins, and sugars.

The achievement opens the way to virtually limitless reproduction of identical progeny from specific genetic material taken from desirable trees. Research is continuing on methods that assure maintenance of genetic stability of the tissue through the growth sequences, and to stimulate the further development of plantlets into normal trees.

Control of Wood Formation

The possibility that wood formation can be manipulated by management of nutrients and hormones normally supplied by the leaves was demonstrated in experiments with decapitated rooted sprouts. Response wood was divergent from the normal in cell type, arrangement, and cell wall thickness. The nutrients provided were undoubtedly supplemented by material emanating from the roots. It is planned to extract such material from roots and attempt to isolate active growth-controlling factors. Kinds, amounts, and interactions of nutrients necessary to initiate wood with normal characteristics in rooted but decapitated sprouts will also be studied.

32

Thinning Improves Loblolly Pine

Loblolly pine trees remaining after a young stand has been thinned can produce improved wood in subsequent years, as judged by increase in specific gravity. This was shown in a study of young loblolly conducted in cooperation with the Crossett Research Center, Crossett, Ark. The finding sheds new light on a problem that long has concerned plantation managers—whether the stimulated growth resulting from release by thinning produces weaker or stronger wood.

Walnut Color Control

Evidence that color of black walnut—a key quality criterion in many uses for this valuable species—can be manipulated by control of soil properties was gained in studies significantly relating certain color properties to soil constituents.

The properties studied were determined spectrophotometrically. They are luminance (lightness or brightness), dominant wavelength (principal hue), and purity (percentage of principal hue). Soil properties included were all major nutrients, soil texture and acidity, and depth of mottling or depth to impervious layer. The spectrophotometric color measurements were in turn related to minimum visible differences in brightness, hue, and purity of hue.

Black walnut specimens used had been obtained from Indiana and Missouri in cooperation with the North Central Forest Experiment Station. The work is one aspect of a three-part study covering many aspects of walnut wood quality.

Among significant relations found between wood color and soil properties was the effect of soil acidity and available phosphorus on luminance. Acidity significantly affected the dominant wavelength, or hue. Visible changes in color also were found with variations in calcium and magnesium. Only calcium was found to change purity of hue enough to be visible.

Stain, Rancid Odors in Oak

The socalled black stain that occurs in living oak trees and down-grades wood of otherwise high quality for flooring and other uses may be related to micro-organisms that develop in host trees under soil moisture conditions that are known to bear on stain formation. Investigation also indicates that bacteria or the combined action of bacteria and staining fungi may cause heartwood boundary stains in oaks growing on certain sites.

Studies of cultures of bacteria taken from living trees indicate that bacteria may initiate conditions within the tree that later may result in emission of rancid odors. Volatile fatty acids of rancid oak were quantitatively and qualitatively analyzed by gas chromatography in the search for an explanation of the formation of both wetwood and rancid odors. University of Wisconsin pathologists and bacteriologists aided in the laboratory work.

Cottonwood Tension Wood

Tension wood of eastern cottonwood differs from normal wood in other respects than the presence of gelatinous fibers, a study of anatomical features showed. Specimens of both types of wood were examined for number of annual growth rings per inch of diameter and average ring width, number of vessels and relative vessel size, and proportion of area occupied by gelatinous fibers, nongelatinous fibers, vessels, and parenchyma cells.

Increases in specific gravity were shown to be associated with increases in proportion of area occupied by gelatinous fibers and with decreases in relative vessel size. These associations are probably explained by the gelatinous extra cell wall layer in gelatinous fibers and by the higher amount of cell wall material per unit area contained in wood with vessels of small diameter.

Peruvian Woods

A 7-year program of identifying and evaluating Peruvian woods brought to light 12 hitherto unknown tree species and 14 species unknown in Peru but found earlier in the adjacent Brazilian Amazon Valley. Also obtained were some interesting findings with regard to the specific gravity and shrinkage characteristics of these tropical woods. A tremendous amount of herbarium material was obtained. FPL's Dr. B. F. Kukachka collaborated with Dr. Louis O. Williams,. curator of Chicago's Field Museum, in the identification work. The study was conducted under Public Law 480, using counterpart funds from the sale of surplus agricultural commodities to finance collections in Peru in cooperation with that Nation's Forest Service.

Mount Vernon's Buckeyes

An FPL technologist's examination of wood from two buckeye trees growing on the grounds of Mount Vernon, George Washington's Potomac River estate, apparently has exploded another myth involving the Father of His Country.

Just as the story of the cherry tree has long been generally discredited, so now is an anecdote that Washington in later life planted seeds from which grew the buckeyes still flourishing on either side of a bowling green on the estate. The story of the buckeyes was chronicled by Charles Sprague Sargent, one-time director of Harvard University's Arnold Arboretum. In 1927, Sargent wrote that Washington gathered the seeds in 1784 and planted them the following spring.

Ring counts by FPL Technologist Robert R. Maeglin on specimens sent to FPL by Robert B. Fisher, estate horticulturist, gave the true age of the trees as 162 and 151 years. If Washington had planted the trees, the ring counts should have totaled 183 years, Maeglin points out.

INFORMATION ACTIVITIES

During 1968, FPL research yielded a crop of 101 technical and scientific publications. The information harvest was made available to the public in 40 Forest Service Research Papers, 58 articles and papers published in scientific, technical and trade journals, and three contributions to books.

Visitors, Consulting and Casual

Consulting visitors—men and women with business and professional interests in wood research—totaled 3,349. They came from 48 States and 1 territory—Nevada and Utah were the only States not represented—as well as 52 foreign countries.

Foreign Nations and the number of visitors they sent were: Argentina 4, Australia 9, Austria 1, Belgium 2, Brazil 5, British Honduras 1, Cameroon 3, Canada 49, Chile 2, Colombia 1, Costa Rica 1, Czechoslovakia 6, Denmark 2, El Salvador 1, England 12, Ethiopia 1, Finland 6, France 1, Germany 10, Greece 1, Guyana 2, Holland 4, Honduras 1, Iceland 1, India 3, Indonesia 3, Israel 3, Italy 2, Japan 19, Kenya 1, Mexico 31, Nepal 1, New Guinea 1, New Zealand 7, Nigeria 2, Norway 3, Pakistan 1, Paraguay 1, Peru 5, Philippines 2, Poland 1, Republic of China 6, Republic of South Africa 4, Scotland 2, Spain 1, Sudan 1, Sweden 12, Switzerland 2, Thailand 3, Turkey 2, Venezuela 1, Zambia 7.

A total of 11,150 general interest visitors came to FPL in 1968.

Mrs. Geraldine Martens, shown weighing a kiln sample under tutelage of PPL seasoning specialist Raymond C. Rietz, was the first woman to enroll in FPL's kiln drying demonstrations since their inception in 1919. Wife of a Richland Center, Wis., wood products manufacturer, she is often called on to operate the dry kilns in her husband's absence.

M 134 483

Exhibits

FPL research on black walnut, the most highly prized fine hardwood native to the United States, was featured in an exhibit shown at the International Woodworking Machinery and Furniture Supply Fair September 14-18 at Louisville, Ky. Highlights of the exhibit were furniture stabilized with polyethylene glycol, a bulking agent that markedly re-

34

Larry Hanshaw, senior at Tougaloo College, Miss., examines strips of paper containing deposits of various components of a xylan hydrolyzate as a summer trainee in wood chemistry research. His FPL mentor, Chemist Ralph Scott, right, coached him in the use of paper chromatography for identifying unknowns.

M 135 314-2

duces swelling and shrinking of wood when diffused into the cell walls; machining experiments on black walnut, and research on the relation of soil nutrients and other factors to walnut color.

Educational Activities

The 93rd Kiln Drying Demonstration brought 29 kiln operators, plant managers, and others from various parts of the United States. The Demonstration was distinguished by the first feminine enrollee, Mrs. Geraldine Martens of Richland Center, Wis. More than 2,300 have enrolled in drying demonstrations dating back to 1919.

Ten student trainees spent the summer at FPL on research assignments directly related to their college work. They came from Massachusetts Institute of Technology, Cambridge; Tougaloo College, Miss.; Dillard University, New Orleans; Louisiana Technical University, Ruston; University of Minnesota at

Morris; Oberlin College, Ohio; University of Montana, Missoula; Viterbo College, LaCrosse, Wis.; Dartmouth College, Hanover, N. H.; and the University of Wisconsin. The student trainee program is conducted annually to provide summer jobs for capable college students and to interest them in careers in forest products research.

Two employees received assistance under the Federal Employee Training Act to do advanced study, one at the New York State College of Forestry at Syracuse University, and the other at the University of Wisconsin.

Whitten Act funds were administered by FPL to support graduate research at four universities, the University of British Columbia, Vancouver, B.C., Colorado State University, Fort Collins, North Carolina State University, Raleigh, and the University of Washington, Seattle..

FPL PUBLICATIONS ISSUED IN 1968

1. Anderson, L.O.
   1968. Construction of Nu-frame research house (Utilizing new wood-frame system). U.S. Forest Serv. Res. Paper FPL 88. Mar.
2. Baechler, R.H.
   1968. Further thoughts regarding variable performance of creosoted marine piling. AWPA Proc. 1968.
3. ——
   1968. Preservative treatments, species characteristics and desired retentions in poles. Proc., Wood Pole Institute, June 17-19, 1968. Colorado State Univ., Fort Collins, Colo. pp. 93-99.
4. Beall, F.C.
   1968. Specific heat of wood—further research required to obtain meaningful data. U.S. Forest Serv. Res. Note FPL-0184. Feb.
5. Bendtsen, B.A.
   1968. Mechanical properties and specific gravity of a randomly selected sample of spruce pine. U.S. Forest Serv. Res. Paper FPL 92. May.
6. ——, and Eslyn, W.E.
   1968. House foundation of treated wood after 30 years service. U.S. Forest Serv. Res. Paper FPL 98. Aug.
7. Blew, J.O., Roth, H.G., and Davidson, H.L.
   1968. Retention and distribution of water-borne preservative in redwood treated at different moisture levels. AWPA Proc. 1968.
8. Bohannan, Billy
   1968. Structural engineering research in wood. J. of the Structural Div., Proc. of ASCE 94(ST2):403-416, Feb.
9. Bulgrin, Erwin H., and Ward, James C.
   1968. Factors contributing to heartwood-boundary stain in living oak. Wood Science 1(1):58-64. July.
10. Caulfield, Daniel F.
    1968. On questioning the X-ray evidence of crystallizability of xylan in situ. Tappi 51(8):371-372. Aug.
11. ——, Yao, Yung-Fang, and Ullman, Robert
    1968. The analysis of low-angle light scattering from simple mixtures. In "X-Ray and Electron Methods of Analysis." Plenum Press. pp. 127-161.
12. Chidester, Gardner H.
    1968. Fiber for tomorrow's requirements. Tappi 51(8):46A-47A. Aug.
13. Clark, Ira T.
    1968. Gas chromatographic analysis of phenols from lignin. J. of Chromatography 6:53-55. Jan.
14. ——, and Green, Jesse
    1968. Production of phenols by cooking kraft lignin in alkaline solutions. Tappi 51:44-48. Jan.
15. Comstock, Gilbert L.
    1968. The relationships between permeability of green and dry eastern hemlock. Forest Prod. J. 18(8):20-23. Aug.
16. Cramer, Calvin O.
    1968. Load distribution in multiple-bolt tension joints. J. of Structural Div., Proc. of ASCE ST5, 1101-1117. May.
17. Doyle, D.V.
    1968. Properties of No. 2 Dense kiln-dried southern pine dimension lumber. U.S. Forest Serv. Res. Paper FPL 96. July.
18. Duff, John E.
    1968. Moisture distribution in wood-frame walls in winter. Forest Prod. J. 18(1):61-64. Jan.
19. Eickner, Herbert W.
    1968. Wood can protect you from fire. U.S. Dept. of Agr. Yearbook of Agriculture. Science for better living, pp. 321-324.
20. Esenther, G.R., and Gray, D.E.
    1968. Subterranean termite studies in southern Ontario. The Canadian Entomologist. 100(8):827-834. Aug.
21. Fahey, D.J.
    1968. Application of chemicals to wet webs of paper and linerboard using the smoothing press. Indian Pulp & Paper 23(1):85-92. July.
22. Fahey, D.J., and Laundrie, J.F.
    1968. Kraft pulps, papers, and linerboard from southern pine thinnings. U.S. Forest Serv. Res. Note FPL-0182. Jan.
23. Feather, Milton S., and Harris, J.F.
    1968. The absence of proton exchange during the conversion of hexose to 5-(hydroxymethyl)-2-furaldehyde. Tetrahedron Letters No. 55, pp. 5807-5810.
24. Fleischer, Herbert O.
    1968. Utilizing all species and all of the tree. Pulp & Paper 42(14):28-30, 32. Apr. 1. Also The Northern Logger and Timber Processor 17(2):18,64-65,74. Aug.
25. Foulger, A.N.
    1968. Effect of aphid infestation on properties of grand fir. Forest Prod. J. 18(1):43-47. Jan.
26. ——, and Hacskaylo, J.
    1968. Stem anatomy variation in cottonwood growing under nutrient-deficient conditions. Proc. of the Eighth Lake States Forest Tree Improvement Conf. Sept. 12-13, 1967. U.S. Forest Serv. Res. Paper NC-23, 1968, pp. 41-47. North Central Forest Exp. Sta., St. Paul, Minn.
27. Fracheboud, M., Rowe, J.W., Scott, R.W., Fanega, S.M., Buhl, A.J., and Toda, J.K. 1968. New sesquiterpenes from the yellow wood of slippery elm. Forest Prod. J. 18(2):37-40. Feb.
28. Gerhards, C.C.
    1968. Effects of type of testing equipment and specimen size on toughness of wood. U.S. Forest Serv. Res. Paper FPL 97. July.
29. ——
    1968. Seasoning factors for modulus of elasticity and modulus of rupture of 4-inch lumber. Forest Prod. J. 18(11):27-35. Nov.
30. Gillespie, Robert H.
    1968. Parameters for determining heat and moisture resistance of a urea-resin in plywood joints. Forest Prod. J. 18(8):35-41. Aug.
31. Gjovik, L.R., and Baechler, R.H.
    1968. Field tests on wood dethiaminized for protection against decay. Forest Prod. J. 18(1):25-27. Jan.
32. Godshall, W.D.
    1968. Effects of vertical dynamic loading on corrugated fiberboard containers. U.S. Forest Serv. Res. Paper FPL 94. July.
33. Hallock, Hiram
    1968. Observations on forms of juvenile core in loblolly pine. U.S. Forest Serv. Res. Note FPL-0188. Feb.
34. ——
    1968. The "taper-tension" saw — a new reduced kerf saw. U.S. Forest Serv. Res. Note FPL-0185. June.
35. Heebink, B.G., and Hefty, F.V.
    1968. Steam post-treatments to reduce thickness swelling of particleboard (Exploratory study). U.S. Forest Serv. Res. Note FPL-0187. Mar.
36. Hefty, F.V., and Brooks, J.K.
    1968. Portable apparatus for measuring surface irregularities in panel products. U.S. Forest Serv. Res. Note FPL-0192. May.

35

37. Hiller, Charlotte H.
1968. Trends of fibril angle variation in white ash. U.S.D.A. Forest Serv. Res. Paper FPL 99. Sept.
38. Hittmeier, M.E., Comstock, G.L., and Hann, R.A.
1968. Press drying nine species of wood. Forest Prod. J. 18(9):91-96. Sept.
39. Horn, R.A., and Simmonds, F.A.
1968. Microscopical and other fiber characteristics of high-yield sodium bisulfite pulps from balsam fir. Tappi 51(1):67A-73A. Jan.
40. James, William L.
1968. Effect of temperature on readings of electric moisture meters. Forest Prod. J. 18(10):23-31. Oct.
41. ————.
1968. Static and dynamic strength and elastic properties of ponderosa and loblolly pine. Wood Science 1(1):15-22. July.
42. Jokerst, R.W.
1968. Long term durability of laboratory-made Douglas-fir flakeboard. U.S. Forest Serv. Res. Note FPL-0199. July.
43. Jordan, C.A.
1968. Container effects in cushioned packages: urethane foam cushioning applied as side pads. U.S. Forest Serv. Res. Paper FPL 91. Apr.
44. Keller, E.L., and Fahey, D.J.
1968. Magnesium bisulfite pulping and papermaking with southern pine. Tappi 51(2):98-103. Feb.
45. Kimball, K.E.
1968. Accelerated methods of drying thick-sliced and thin-sawed loblolly pine. Forest Prod. J. 18(1):31-38. Jan.
46. Kirk, T. Kent, Harkin, John M., and Cowling, Ellis B.
1968. Degradation of the lignin model compound syringylglycol-B-guaiacyl ether by Polyporus versicolor and Stereum frustulatum. Biochimica et Biophysica Acta 165(1968):145-163.
47. ————, Harkin, John M., and Cowling, Ellis B.
1968. Oxidation of guaiacyl- and veratryl-glycerol-B-guaiacyl ether by Polyporus versicolor and Stereum frustulatum. Biochimica et Biophysica Act 165 (1968):134-144.
48. Kurtenacker, R.S., and Godshall, W.D.
1968. Performance of nailed cleats in blocking and bracing applications. U.S.D.A. Forest Serv. Res. Note FPL-0200. Aug.
49. Lewis, Wayne C.
1968. Hardness modulus as an alternate measure of hardness to the standard Janka ball for wood and wood-base materials. U.S. Forest Serv. Res. Note FPL-0189. Mar.
50. ————.
1968. Thermal insulation from wood for buildings: Effects of moisture and its control. U.S. Forest Serv. Res. Paper FPL 86. July.
51. McMillen, John M.
1968. Prevention of pinkish-brown discoloration in drying maple sapwood. U.S. Forest Serv. Res. Note FPL-0193. May.
52. ————.
1968. Transverse strains during drying of 2-inch ponderosa pine. U.S. Forest Serv. Res. Paper FPL 83. June.
53. Maeglin, R.R., Nelson, N.O., and Wahlgren, H.E.
1968. Walnut wood characteristics in relation to soil-site conditions. Proc. of the Eighth Lake States Forest Tree Improvement Conf. Sept. 12-13, 1967. U.S. Forest Serv. Res. Paper NC-23, 1968, pp. 37-40. North Central Forest Exp. Sta., St. Paul, Minn.
54. Maki, A.C.
1968. Finite element techniques for orthotropic plane stress and orthotropic plate analysis. U.S. Forest Serv. Res. Paper FPL 87. June.
55. Malcolm, F.B.
1968. Sacrificing short butt log to chips may reduce lodgepole stud warping. Forest Industries 95(5):88-89. May.
56. ————.
1968. Warp in studs from small-diameter loblolly pine. South. Lbrmn. 216(2687):27-30. Apr.
57. Mitchell, Harold L.
1968. PEG is the sweetheart of the wood craftsman. U.S. Dept. of Agr. Yearbook 1968: Science for better living, pp. 147-149.
58. Mohaupt, A.A.
1968. Tests give film barriers the air, show how well they do. Package Engineering 13(12):80-84. Dec.
59. Moore, W.E., Effland, M.J., and Roth, H.G.
1968. Detection of petroleum oil diluents in coal tar creosote by thin layer chromatography. J. of Chromatography 35(4):522-525, Dec. 17.
60. Munthe, B.P., and Ethington, R.L.
1968. Method for evaluating shear properties of wood. U.S. Forest Serv. Res. Note FPL-0195. June.
61. Orosz, Ivan
1968. Some nondestructive parameters for prediction of strength of structural lumber. U.S.D.A. Forest Serv. Res. Paper FPL 100. Oct.
62. Panek, Edw.
1968. Study of paintability and cleanliness of wood pressure treated with water-repellent preservative. AWPA Proc. 1968.
63. Peters, Curtis C.
1968. Multiple-flitch method for thick slicing. Forest Prod. J. 18(9):82-83. Sept.
64. ————, and Zenk, R.R.
1968. Effect of precompression on sliced wood ½ and 1 inch in thickness. U.S. Forest Serv. Res. Note FPL-0194. July.
65. Peters, C.C., Zenk, R.R., and Mergen, A.
1968. Effects of roller-bar compression and restraint in slicing wood 1 inch thick. Forest Prod. J. 18(1):75-80. Jan.
66. Pronin, Dimitri, and Vaughan, Coleman L.
1968. A literature survey of Populus species with emphasis on P. tremuloides. U.S. Forest Serv. Res. Note FPL-0180. Rev. Aug.
67. Quirk, John T.
1968. Fluorescence microscopy for detecting adhesives on fracture surfaces. U.S. Forest Serv. Res. Note FPL-0191. Apr.

68. ————, Kozlowski, T.T., and Blomquist, R.F.
1968. Contribution of end-wall and lumen bonding to strength of butt joints. U.S. Forest Serv. Res. Note FPL-0179. Jan.
69. Rice, James W., and Sanyer, Neomi
1968. Sodium sulfite recovery by the direct oxidation of smelt. Tappi 51(7):321-327. July.
70. Rietz, R.C., and Jenson, J.A.
1968. Producing check-free beech for turnings. Forest Prod. J. 18(11):42-44. Nov.
71. Sanyer, Necmi
1968. Progress and prospects of polysulfide pulping. Tappi 51(8):48A-51A. Aug.
72. Schaeffer, R.E.
1968. Gluing fire-retardant-treated Douglas-fir and western hemlock. U.S.D.A. Forest Serv. Res. Note FPL-0160. Nov.
73. Schaffer, E.L.
1968. A simplified test for adhesive behavior in wood sections exposed to fire. U.S. Forest Serv. Res. Note FPL-0175. Nov.
74. Seikel, M.K., Hostettler, F.D., and Johnson, D.B.
1968. Lignans of Ulmus thomasii heartwood-I. Thomasic Acid. Tetrahedron Vol. 24, pp. 1475-1488.
75. Setterholm, Vance C., Benson, Roy, and Kuenzi, Edward W.
1968. Method for measuring edgewise shear properties of paper. Tappi 51(5):196-202. May.
76. Simpson, William T., and Skaar, Christen
1968. Effect of restrained swelling on wood moisture content. U.S. Forest Serv. Res. Note FPL-0196. July.
77. ————, and Skaar, Christen
1968. Effect of transverse compressive stress on loss of wood moisture. U.S. Forest Serv. Res. Note FPL-0197. July.
78. ————, and Soper, Vernon R.
1968. Stress-strain behavior of films of four adhesives used with wood. U.S. Forest Serv. Res. Note FPL-0198. July.
79. Skaar, C., and Simpson, William
1968. Thermodynamics of water sorption by wood. Forest Prod. J. 18(7):49-58. July.
80. Smith, Diana M.
1968. Wood quality of loblolly pine after thinning. U.S. Forest Serv. Res. Paper FPL 89. May.
81. Springer, Edward L., and Zoch, Lawrence L.
1968. Hydrolysis of xylan in different species of hardwoods. Tappi 51(5):214-218. May.
82. Steinmetz, P.E., and Fahey, D.J.
1968. Resin treatments for improving dimensional stability of structural fiberboard. Forest Prod. J. 18(9):82-83. Sept.
83. Stern, R.K.
1968. Flat-crush cushioning capability of corrugated fiberboard pads under repeated loading. U.S. Forest Serv. Res. Note FPL-0183. Feb.
84. ————.
1968. Tests show corrugated pads' performance as cushioning. Package Engineering. Feb.
85. Tang, Walter K., and Eickner, Herbert W.
1968. Effect of inorganic salts on pyrolysis of wood, cellulose, and lignin determined by differential thermal anaylsis. U.S. Forest Serv. Res. Paper FPL 82. Jan.
86. Tarkow, Harold, and Feist, W.C.
1968. The superswollen state of wood. Tappi 51(2):80-83. Feb.
87. ————, and Seborg, Raymond
1968. Surface densification of wood. Forest Prod. J. 18(9):104-7. Sept.
88. U.S. Forest Products Laboratory
1968. Water-repellent preservatives. U.S. Forest Serv. Res. Note FPL-0124, Aug.
89. Uyeo, S., Okada, J., Matsunaga, S., and Rowe, J.W.
1968. The structure and the stereochemistry of abieslactone. Tetrahedron Vol. 24 (2859-2880).
90. Wahlgren, Harold E., Baker, Gregory, Maeglin, Robert R., and Hart, Arthur C. 1968. Survey of specific gravity of eight Maine conifers. U.S. Forest Serv. Res. Paper FPL 95. July.
91. Weatherwax, R.C., and Tarkow, Harold
1968. Cell wall density of dry wood. Forest Prod. J. 18(2):83-85. Feb.
92. ————, and Tarkow, Harold
1968. Importance of penetration and adsorption compression of the displacement fluid. Forest Prod. J. 18(7):44-46. July.
93. Wengert, E.M.
1968. Electrical analog approach to heat flow through wood-frame walls. Forest Prod. J. 18(1):99-101. Jan.
94. ————, and Koster, A.L.
1968. Sun-following rack accelerates weathering of wood products. Solar Energy 12:267-272.
95. Wilcox, W. Wayne
1968. Changes in wood microstructure through progressive stages of decay. U.S. Forest Serv. Res. Paper FPL 70. July.
96. Wilkinson, T.L.
1968. Longtime performance of trussed rafters: initial evaluation. U.S. Forest Serv. Res. Paper FPL 93. July.
97. ————.
1968. Strength evaluation of round timber piles. U.S.D.A. Forest Serv. Res. Paper FPL 101. Dec.
98. Wolter, Karl E.
1968. A new method for marking xylem growth. Forest Science 14(1):102-104. Mar.
99. ————.
1968. Root and shoot initiation in aspen callus cultures. Nature 219(5153):509-510. Aug.
100. Zinkel, D.F., and Zank, L.C.
1968. Separation of resin from fatty acid methyl esters by gel-permeation chromatography. Analyt. Chem. 40:1144-1146. June.
101. ————, Lathrop, Mary B., and Zank, L.C.
1968. Preparation and gas chromatography of the trimethylsilyl derivatives of resin acids and the corresponding alcohols. J. of Gas Chromatography 6(3):158-160. Mar.